The Atmosphere and Ocean

Frontispiece Major topographic features of the ocean and the continents

Land above 2000 metres

Continental Shelf (200 metres)

200–4000 metres

> 4000 metres

Major trench

R

The Atmosphere and Ocean

A PHYSICAL INTRODUCTION
SECOND EDITION

Neil Wells

Department of Oceanography, University of Southampton, UK

JOHN WILEY & SONS

Chichester · New York · Weinheim · Brisbane · Singapore · Toronto

Other Wiley Editorial Offices

John Wiley & Sons, Inc., 605 Third Avenue,
New York, NY 10158-0012, USA

VCH Verlagsgesellschaft mbH, Pappelallee 3,
D-69469 Weinheim, Germany

Jacaranda Wiley Ltd, 33 Park Road, Milton,
Queensland 4064, Australia

John Wiley & Sons (Asia) Pte Ltd, 2 Clementi Loop #02-01,
Jin Xing Distipak, Singapore 129809

John Wiley & Sons (Canada) Ltd, 22 Worcester Road,
Rexdale, Ontario M9W 1L1, Canada

Library of Congress Cataloging-in-Publication Data
Wells, Neil.
    The Atmosphere and the Ocean: a physical introduction / by Neil Wells. — 2nd ed.
        p.   cm.
    Includes bibliographical references and index.
    ISBN 0-471-96216-3
    1. Atmospheric physics.   2. Oceanography.   I. Title.
    QC861.2.W45   1997
    551.5–dc21   96-48904
                                                                 CIP
British Library Cataloguing in Publication Data

A catalogue record for this book is available from the British Library

ISBN 0-471-96216-3

Typeset in 11/13pt Palatino by MHL Typesetting Ltd, Coventry
Printed and bound in Great Britain by Biddles Ltd, Guildford and King's Lynn
This book is printed on acid-free paper responsibly manufactured from sustainable forestation,
for which at least two trees are planted for each one used for paper production.

Contents

Preface to the Second Edition

The second edition of *The Atmosphere and Ocean* has been prompted by the need to update the material. The body of knowledge in this area has grown substantially over the last decade, as has the data available. The approach used has been conservative in that changes to the text have been introduced only where the original material has been superseded by new observations and interpretations. For example, the latter part of Chapter 10 has been rewritten to take into account the growth in awareness of global climate change and the complexities and fascinations of this relatively new area of science.

In the last decade we have seen the completion of the Tropical Ocean Global Atmosphere (TOGA) programme of which the TOGA-COARE experiment in 1991–92 in the western equatorial Pacific Ocean was the centrepiece. This experiment involved international meteorologists and oceanographers in an intensive observation period of the tropical troposphere and the upper ocean. It assisted in the development of current understanding of atmospheric and ocean processes and their interactions.

The World Ocean Circulation Experiment (WOCE), from 1990 to 1997, has provided a global snapshot of the state of the oceans by providing data of unprecedented accuracy and sampling. More observations have been collected in this survey than in all previous oceanography programmes. This unique data set will provide scientists with a benchmark from which global ocean change can be measured. It will also provide material for ocean scientists to test their ideas and their mathematical models of ocean circulation.

I hope that I have managed to convey some of the enthusiasm and fascination that I have for the subject and that this book will provoke similar reactions in my readers.

I wish to thank, Rachel Mills for her invaluable help on the early history of the earth, Kate Davis for the drafting of all the figures and my colleagues and students in the Department of Oceanography at the University of Southampton for their interest and help. Particular thanks go to Ian Robinson who, whilst Head of Department, supported my visit to the Flinders University of South Australia where a large part of the book was revised. The patience of Amanda Hewes and her colleagues at John Wiley, as deadlines came and went, is gratefully acknowledged. Finally I owe a considerable debt of gratitude to Jenny, Mark and Helen who allowed me to indulge my interest at weekends and in the evenings over the past three years.

1

The Earth within the Solar System

1.1 The Sun and its constancy

Any account of the Earth's atmosphere and ocean cannot be regarded as complete without a discussion of the Sun, the solar system and the place of the Earth within this system. The Sun supplies the energy absorbed by the Earth's atmosphere–ocean system. Some of its energy is converted directly into thermal energy, which drives the atmospheric circulation. A small portion of this energy appears as the kinetic energy of the winds which, in turn, drives the ocean circulation. Some of the intercepted solar energy is transformed by photosynthesis into biomass, a large proportion of which is ultimately converted into heat energy by chemical oxidation within the bodies of animals and by the decomposition and burning of vegetable matter. A very small proportion of the photosynthetic process produces organic sediments, which may eventually be transformed into fossil fuels. It is estimated that the solar radiation intercepted by the Earth in 10 days is equivalent to the heat which would be released by the combustion of all known reserves of fossil fuels on Earth. The Sun is therefore of fundamental importance in the understanding of the uniqueness of the Earth.

The Sun is a main sequence star in the middle stages of its life and was formed 4.6×10^9 years ago. It is composed mainly of hydrogen (74% by mass) and helium (25% by mass); the remaining 1% of the Sun's mass comprises the elements oxygen, nitrogen, carbon, silicon, iron, magnesium and calcium. The emitted energy of the Sun is 3.8×10^{26} W and this energy emission arises from the thermonuclear fusion of hydrogen into helium at temperatures around 1.5×10^6 K in the core of the Sun.

In the core, the dominant constituent is helium (65% by mass) and the hydrogen content is reduced to 35% by mass as a direct result of its consumption in the fusion reactions. It is estimated that the remaining hydrogen in the Sun's core is sufficient to maintain the Sun at its present luminosity and size for a further 4×10^9 years. At this stage it is expected that the Sun will expand into a red giant and engulf all of the inner planets of the solar system (i.e. Mercury, Venus, Earth and Mars).

There exists a high-pressure gradient between the core of the Sun and its perimeter, and this is balanced by the gravitational attraction of the mass of the Sun. In the core, the energy released by the thermonuclear reaction is transported by energetic photons but, because of the strong absorption by the peripheral gases, most of these photons do not penetrate the surface. This absorption causes heating in the region outside the core. In contrast, the outer layers of the Sun are continually losing energy by radiative emission into space in all regions of the electromagnetic spectrum. This causes a large temperature gradient to develop between the surface and the inner region of the Sun. This large temperature gradient produces an unstable region and large-scale convection currents are set up which transfer heat to the surface of the Sun. These convection currents are visible as the fine grain structure, or granules, in high-resolution photographs of the surface of the Sun. It is thought that the convection currents have a three-tier structure within the Sun. The largest cells, 200×10^3 km in diameter, are close to the core. In the middle tier the convection cells are about 30×10^3 km in diameter, and at the surface they are 10^3 km. The latter cells have a depth of 2000 km. In each cell, hot gas is transported towards the cooler surface, whilst the return flow transports cooler gas towards the interior.

Almost all of the solar radiation emitted into space, approximately 99.9%, originates from the visible disc of the Sun, known as the photosphere. The photosphere is the region of the Sun where the density of the solar gas is sufficient to produce and emit a large number of photons, but where the density of the overlying layers of gas is sufficient to absorb the emitted photons. This region of the Sun has a thickness of approximately 500 km, but no sharp boundaries can be defined. The radiation spectrum of the Sun, when fitted to a blackbody curve, gives a blackbody temperature of 6000 K, although the effective temperature, deduced from the total energy emitted by the Sun, i.e. its luminosity, gives a lower temperature of 5800 K.

The photosphere is not uniform in temperature. The lower regions of the photosphere have temperatures of 8000 K, whilst the outer regions have temperatures of 4000 K. Furthermore, the convection cells produce horizontal temperature variations of 100 K between the ascending and descending currents of solar gas. Larger convection cells also appear in the photosphere and they have diameters of 30 000 km. They appear to originate from the second tier of convection within the Sun. The appearance of sunspots give rise to horizontal variations of 2000 K within the photosphere. The inner regions of the sunspot have blackbody temperatures of 4000 K. Sunspots have diameters of 10 000–150 000 km and they may last for many weeks. It is thought that the 'sunspot' causes a localized suppression of convection and therefore leads to a reduction in the transfer of heat into the photosphere from the interior. However, although the sunspot features are dramatic, they occupy less than 1% of the Sun's disc, and therefore the effect on the luminosity of the Sun is small.

Beyond the photosphere lies the chromosphere where the temperature decreases to a minimum of 4000 K at 2000 km above the photosphere and then increases sharply to a temperature of 10^6 K at a height of 5000 km in the region of the corona. However, because of the low density of the gas in this region, the radiation emitted from the chromosphere and the corona only amount to 0.1% of the total radiation from the Sun.

These different temperature zones of the Sun can also be observed in the solar spectrum (Figure 1.1). The visible and infra-red radiation emitted from the photosphere follow reasonably closely the blackbody curve for a temperature of 5800 K. However, substantial deviations from the theoretical curve occur in the X-ray and radio wavebands, and lesser deviations occur in the ultraviolet spectrum. The high temperature of the corona is responsible for the intense X-ray band, whilst the high radio frequency energy is associated with the solar wind and solar activity. However, the energy in these wavebands is a negligible fraction of the total emitted energy and therefore these very variable regions of the spectrum have little direct influence on the total solar energy received on the Earth. The depletion of energy in the ultraviolet spectrum is the result of emission at a lower temperature than the photosphere and therefore probably originates in the temperature minimum of the lower chromosphere. Recent observations have shown that the ultraviolet

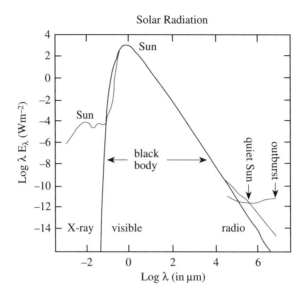

Figure 1.1 Solar spectrum and blackbody curve: energy distribution of the Sun and a black body at 5800 K

energy is not constant and shows considerable variability in the short-wave part of the spectrum, amounting to 25% of its average value, at 0.15 μm, and 1% at 0.23 μm. Again, the amount of energy in this band is relatively small and contributes less than 0.1% to the total energy. Furthermore, satellite observations of the solar spectrum since 1978 have demonstrated a 0.1% variation of the solar output over the 11-year sunspot cycle. These variations are rather small when compared to those produced by orbital variations of the Earth around the Sun.

1.2   Orbital variations in solar radiation

Let S be the total solar output of radiation in all frequencies. At a distance, r, imagine a sphere of radius r on which the flux of radiation will be the same (assuming that the radiation from the Sun is equal in all directions). If the flux of radiation per unit area at distance r is given by $Q(r)$, then the total radiation on the imagined sphere is $4\pi r^2 Q(r)$.

In the absence of additional energy sources to the Sun:

$$S = 4\pi r^2 Q(r)$$

Rearranging:

$$Q(r) = \frac{S}{4\pi r^2} \tag{1.1}$$

In practice, the solar radiation cannot be measured at the Sun, but it can be measured by satellites above the Earth's atmosphere. Recent determinations of the flux of radiation per unit area, Q, give a value of $1360\,\mathrm{W\,m^{-2}}$. Given that the Earth is approximately 1.5×10^{11} m away from the Sun, S can then be calculated to be 3.8×10^{26} W.

Though S is the true solar constant, in meteorology Q is defined as the solar constant for the Earth. Table 1.1 shows values of the solar constant obtained for other planets in the solar system. It is noted that the dramatic changes in the solar constant between Earth and our nearest planetary neighbours, Mars and Venus, merely serve to highlight the uniqueness of the position of the Earth in the solar system. At the radius of Pluto (some 39 Earth–Sun distances), the flux of the radiation from the Sun is less than $1\,\mathrm{W\,m^{-2}}$.

The radiation incident on a spherical planet is not equal to the solar constant of that planet. The planet intercepts a disc of radiation of area πa^2, where a is the planetary radius, whereas the surface area of the planet is $4\pi a^2$. Hence, the solar radiation per unit area on a spherical planet is

$$\frac{Q\pi a^2}{4\pi a^2} = \frac{Q}{4} \tag{1.2}$$

Table 1.1 Radiative properties of terrestrial planets. Q is the solar irradiance at distance r from the Sun, α is the planetary albedo, T_e is the radiative equilibrium temperature and T_s is the surface temperature

	r $(10^9\,\mathrm{m})$	Q $(\mathrm{W\,m^{-2}})$	α	T_e (K)	T_s (K)
Venus	108	2623	0.77	227	760
Earth	150	1360	0.30	255	288
Mars	228	589	0.24	211	230

The average radiation on the Earth's surface is $340\,\mathrm{W\,m^{-2}}$. The above discussion assumes the total absence of an atmosphere and that the Earth is a perfect sphere in a spherical orbit.

The three geometrical factors which determine the seasonal variation of solar radiation incident on the Earth are shown in Figure 1.2. The Earth revolves around the Sun in an elliptical orbit, being closest to the Sun, i.e. at perihelion, about 3 January and farthest from the Sun, i.e. at aphelion, on 5 July. The eccentricity, represented by the symbol e, of the present-day orbit is 0.017. At aphelion, it can be shown that

$$r_{\mathrm{aphelion}} = (1 + e)\bar{r}$$

where $\bar{r}$ is the mean distance between the Earth and the Sun.

As the incident radiation is inversely proportional to the square of the Earth's distance from the Sun, it can also be shown that, at aphelion, the radiation received by the Earth will be 3.5% less than the annual average. Furthermore, the total radiation incident on the Earth over the course of one year is independent of the eccentricity of the Earth's orbit. The long-term variation in eccentricity indicates changes of $0.005 - 0.060$ occurring with a period of $100\,000$ years. This would produce changes of the order of $\pm 10\%$ in the radiation incident on the Earth at aphelion and perihelion.

However, the angle of tilt of the Earth with respect to the plane of rotation of the Earth's orbit, i.e. the obliquity of the ecliptic, is the dominant influence on the seasonal cycle of solar radiation (Figure 1.3). It not only determines the march of the seasons, but it is important in determining the latitudinal distribution of the climatic zones. At the present time this angle is 23.5⁰, which is similar to the obliquity of Mars. It can be shown that the amplitude of the seasonal variation in the solar radiation is directly proportional to the obliquity. If the obliquity were reduced, there would be a reduction in the amplitude of the seasonal solar radiation cycle, and if the obliquity were zero, the seasonal variation would depend solely on the eccentricity of the orbit.

The obliquity has varied between 22 and 24.5⁰ over the past million years and this variation has had a period of $41\,000$ years. Such variations cannot influence the total radiation intercepted by the Earth, but they will exert an influence on the seasonal cycle. The sensitivity of the global temperature to the seasonal cycle, which arises from ice,

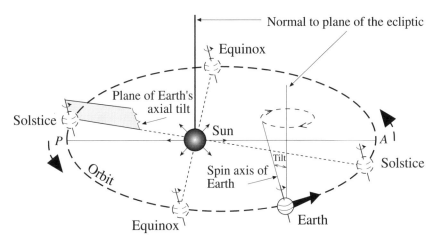

Figure 1.2   Geometry of the Earth–Sun system. The Earth's orbit, the large ellipse with major axis AP and the Sun at one focus, defines the plane of the ecliptic. The plane of the Earth's axial tilt (obliquity) is shown passing through the Sun corresponding to the time of the southern summer solstice. The Earth moves around its orbit in the direction of the solid arrow (period one year) whilst spinning about its axis in the direction shown by the thin curved arrows (period one day). The broken arrows shown opposite the points of aphelion (A) and perihelion (P) indicate the direction of the very slow rotation of the orbit

cloud and ocean feedback processes, indicates that these changes may be sufficient to initiate long-term changes in climate and, in the most extreme cases, ice ages.

The third parameter which affects the seasonal cycle is the longitude of the perihelion, measured relative to the vernal equinox. It precesses by 360º in a period of 21 000 years. Therefore, in about 10 500 years' time, the Earth will be closest to the Sun in July and farthest away in January. Milankovitch, a Serbian scientist, recognized the relationship between the variations in Earth's orbit around the Sun and climate change over the last two million years. In particular, he determined the orbital conditions which would lead to the initiation of an ice age over the Northern Hemisphere continents. He reasoned that a reduction in solar radiation during the polar summer would lead to a reduction in summer melting of the snow cover. This in turn, over many years, would lead to the build-up of ice over the continents. The optimum orbital conditions for this reduction in the summer solar radiation are:

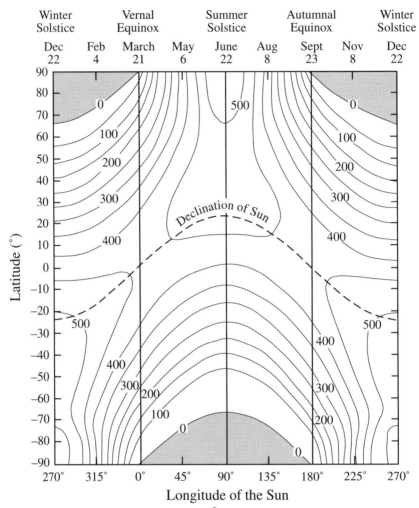

Figure I.3 Solar radiation in W m^{-2} arriving at the Earth's surface, in the absence of an atmosphere, as a function of latitude and time of year

(i)   That aphelion occur during the Northern Hemisphere summer.
(ii) That the eccentricity be large, to maximize the Earth–Sun separation at aphelion.
(iii) That the obliquity be small.

These three factors would act to decrease the amplitude of the seasonal solar radiation cycle, and therefore reduce the solar radiation in summer at high latitudes in the Northern Hemisphere. However, all of these factors would tend to lead to an increase in solar radiation in the Northern Hemisphere winter. Comparison of the orbital parameters with long-term temperature records, deduced from the dating of deep-ocean sediment cores, indicate that there is a good statistical agreement with theory.

1.3   Radiative equilibrium temperature

The inner planets of the solar system have one common attribute – the lack of a major source of internal energy. It is of course true that there is a steady geothermal flow of energy from the interior of the Earth as a result of radioactive decay in the Earth's core, but this energy flow is less than $0.1\,\mathrm{W\,m^{-2}}$ and can be compared with an average solar heat flux of $340\,\mathrm{W\,m^{-2}}$. What, then, is the fate of this steady input of solar energy? On average, about 30% is scattered back into space, by clouds, snow, ice, atmospheric gases and aerosols, and the land surface, leaving 70% available for the heating of the atmosphere, the ocean and the Earth's surface. However, the annual average of the Earth's temperature has not changed by more than $0.5\,\mathrm{K}$ over the past 100 years. Why, then, have the temperatures of the atmosphere and the ocean not increased? The observation that both the temperature of the atmosphere and the ocean have remained relatively constant implies that energy is being lost from the ocean, the atmosphere and the Earth's surface at approximately the same rate as it is being supplied by the Sun. The only mechanism by which this heat can be lost is by the emission of electromagnetic radiation from the Earth's atmosphere into space. If it is assumed that the Earth can be represented as a black body, then it is possible to apply Stefan's law to the electromagnetic emission of the Earth. Stefan's law states that the total emission of radiation from a black body over all wavelengths is proportional to the fourth power of the temperature, i.e. $E = \sigma T^4$, where σ is the Stefan–Boltzmann constant. For the emission by the Earth of an amount of radiation equal to that received from the Sun, there can be defined a radiative equilibrium temperature, T_e.

If $Q/4$ is the energy flux from the Sun and α is the fraction of solar radiation reflected back into space, then for the Earth's system in thermal equilibrium:

$$(Q/4)(1 - \alpha) = \sigma T_e^4 \qquad (1.3)$$

where $\alpha = 5.67 \times 10^{-8}\,\mathrm{W\,m^{-2}\,K^{-4}}$. By inserting $\alpha = 0.3$ and $Q/4 = 340\,\mathrm{W\,m^{-2}}$, a radiative equilibrium temperature of 255 K, or -18°C, can be calculated.

This temperature is rather lower than might be expected. The mean surface temperature of the Earth is 15°C, or approximately 288 K, and so it is clear that the radiative equilibrium temperature bears little direct relationship to the observed surface temperature. This is because most of the planetary radiation is emitted by the atmosphere, whilst only a small fraction originates from the surface of the Earth. The temperature of the atmosphere decreases by $6.5\,\mathrm{K\,km^{-1}}$ from the surface to the tropopause, 10 km above the Earth's surface, and therefore the radiative equilibrium temperature corresponds to the temperature at a height of 5 km.

The Wien displacement equation enables the calculation of the wavelength of maximum radiation, λ_{max}, and it states that:

$$\lambda_{max} T = 2897\,\mu\mathrm{m\,K} \qquad (1.4)$$

If the radiative equilibrium temperature, $T_e = 255$ K, is substituted in this equation, then λ_{max} is calculated to be approximately 11 μm, which lies in the middle part of the infra-red spectrum. In this region of the electromagnetic spectrum, water and carbon dioxide have large absorption bands so that all substances containing water are particularly good absorbers of radiation in the middle infra-red. As far as emission is concerned, water has an emissivity of 0.99, which means that it emits radiation at the rate of 99% of the theoretical blackbody value. Hence clouds, which are composed mainly of water droplets, are good infra-red emitters. The prevalence of liquid water and water vapour in the Earth's atmosphere therefore leads to the strong absorption and re-emission of radiation. This ability of water to absorb and re-emit radiation back to the Earth's surface results in the higher observed mean surface temperature. If the atmosphere was transparent to the emitted planetary radiation, then the surface temperature would be close to the radiative equilibrium temperature. The temperature of the surface of a planet without an atmosphere, such as Mercury, would be observed to have a surface equilibrium temperature equal to that of the radiative equilibrium temperature. Table 1.1 shows the radiative equilibrium temperatures for Mars and Venus, as well as their surface temperatures. The surface temperatures are higher than the radiative temperatures. On Mars the atmospheric mass is smaller than that of the Earth by two orders of magnitude and therefore, though the carbon dioxide absorbs and re-emits planetary radiation back to the surface, a surface warming of only 19 K is produced. However, on Venus the surface temperature of 760 K is sufficient to melt lead, in spite of the fact that the radiative equilibrium temperature is less than that of the Earth. This low radiative equilibrium temperature is caused by the reflection of 77% of the incident solar radiation back into space by the omnipresent clouds, which are not composed of water droplets as they are on Earth. Therefore, though the planet receives less net solar radiation than the Earth, it has a surface temperature 472 K higher. This is the result of the massive carbon dioxide atmosphere, which absorbs virtually all of the radiation emitted by the surface and re-emits it back to the surface. Furthermore, it is known that clouds of sulphuric acid also enhance this warming effect.

It is clear that, although the distance from the Sun determines the energy input into an atmosphere, the mass and the constituents of that atmosphere are also important factors in the determination of the surface temperature of the planet.

1.4   Thermal inertia of the atmosphere

The thermal inertia of the atmosphere gives an indication of how quickly the atmospheric temperature would respond to variations in solar radiation, and it is therefore of importance in the understanding of climatic change. Consider an atmosphere having a mass M, per unit area, and a specific heat at constant pressure, C_p. The thermal inertia of the atmosphere is MC_p.

Initially, the atmosphere is in thermal equilibrium and therefore the solar radiation absorbed by the atmosphere is in balance with the emission of long-wavelength planetary radiation into space (equation 1.3).

Now consider the situation where heat is suddenly added to the atmosphere, for example, as the result of the burning of fossil fuels or by a large thermonuclear explosion. How long would the atmosphere take to regain its former equilibrium? If all the heat were liberated at the same time, then the atmosphere's temperature would increase suddenly by ΔT. The planetary radiation emitted into space would now be $\sigma(T_e + \Delta T)^4$, and therefore more radiation would be emitted than received from the Sun. This deficit implies that the atmosphere would cool. The temperature change, dT/dt, is proportional to the net difference between the emitted planetary radiation and the solar radiation, and thus

$$MC_p \frac{dT}{dt} = -\sigma(T_e + \Delta T)^4 + \frac{Q}{4}(1 - \alpha)$$

Rearranging and substituting from equation 1.3:

$$MC_p \frac{dT}{dt} = -\sigma T_e^4 \left(1 + \frac{\Delta T}{T_e}\right)^4 + \sigma T_e^4$$

Expanding by the binomial expansion:

$$MC_p \frac{dT}{dt} = -\sigma T_e^4 \left(1 + \frac{4\Delta T}{T_e} + 6\left[\frac{\Delta T}{T_e}\right]^2 + 4\left[\frac{\Delta T}{T_e}\right]^3 + \left[\frac{\Delta T}{T_e}\right]^4\right)$$
$$+ \sigma T_e^4$$

Providing that $\Delta T/T_e \ll 1$, then the higher-order terms in the above expression can be neglected. Therefore, retaining only the first-order terms:

$$MC_p \frac{dT}{dt} = -\sigma T_e^4 \frac{4\Delta T}{T_e}$$

Since $T = T_e + \Delta T$, where T_e is independent of time. Then

$$\frac{dT}{dt} = \frac{d\Delta T}{dt}$$

Hence

$$\frac{d}{dt}\Delta T = \left(\frac{4\sigma T_e^3}{MC_p}\right)\Delta T \qquad (1.5)$$

The solution of the above equation is

$$\Delta T(t) = \Delta T_0 \exp(-t/\tau) \qquad (1.6)$$

where ΔT_0 is the initial temperature perturbation at $t = 0$ and τ_R is known as the radiation relaxation time constant:

$$\tau_R = \frac{MC_p}{4\sigma T_e^3} \qquad (1.7)$$

For Earth's atmosphere, where $M = 10.316 \times 10^3\,\text{kg m}^{-2}$, $\sigma = 5.67 \times 10^{-8}\,\text{W m}^{-2}\,\text{K}^{-4}$, $C_p = 1004\,\text{J kg}^{-1}\,\text{K}^{-1}$ and $T_e = 255\,\text{K}$, τ_R can be calculated to be 2.7×10^6 s or 32 days. From equation 1.6:

$$\Delta T = \Delta T_0 e^{-1} \text{ at } t = \tau_R$$
$$\Delta T = \Delta T_0 e^{-2} \text{ at } t = 2\tau_R$$
and
$$\Delta T = \Delta T_0 e^{-3} \text{ at } t = 3\tau_R$$

For an initial temperature perturbation of 1 K, equivalent to the instantaneous burning of 240×10^9 t of coal, then after approximately 32 days, the temperature perturbation will be reduced to 0.37 K; after 64 days it will be 0.13 K and after 96 days it will be 0.05 K. For the Earth's atmosphere, therefore, the colossal excess of heat will be lost by planetary radiation to space within 100 days and the Earth's radiative equilibrium will be restored.

Although this particular problem is hypothetical, the concept is a useful one for understanding the response of the atmosphere to changes in the radiative balance caused by the daily and seasonal variations in solar radiation. For instance, because the radiative time constant is much greater than one day, the lower part of the Earth's atmosphere (wherein is located the major part of the mass of the atmosphere) does not respond significantly to the daily variation in solar radiation. However, for seasonal variations, where the solar radiation changes are of a longer period than the radiative time constant, there will be a significant response in the radiation temperature of the atmosphere.

Table 1.2 shows a comparison of the radiative time constant of Mars, Venus and the Earth. On Mars, because of the small thermal inertia, the diurnal variation of temperature at the surface is very large and, as a result, large thermal tides are set up in the atmosphere. This

Table 1.2 Radiative time constant, τ_R, for the terrestrial planets. C_p is the specific heat at constant pressure, p is the surface pressure and g is the acceleration due to gravity

	C_p $(\mathrm{J\,kg^{-1}\,K^{-1}})$	p (kPa)	g $(\mathrm{m\,s^{-2}})$	τ_R (days)
Venus	830	9000	8.9	3657
Earth	1004	100	9.8	32
Mars	830	0.8	3.7	1

Figure 1.4 Seasonal variation of the zonal mean long-wave radiation flux, $\mathrm{W\,m^{-2}}$, measured by satellite from July 1975 to December 1976. A flux of $260\,\mathrm{W\,m^{-2}}$ corresponds to a radiation temperature of 260 K, whilst a flux of $160\,\mathrm{W\,m^{-2}}$ is equivalent to a temperature of 230 K

phenomenon is also found in the upper reaches of the Earth's atmosphere, where the atmospheric mass is small and the thermal capacity is reduced. However, on Venus the diurnal variation in temperature is only 1–2 K because of the very large radiative time constant, even though the Venusian day is 243 Earth days long!

It must, of course, be remembered that the seasonal change in temperature on Earth is modified by the thermal capacity of the ocean. The thermal capacity of the atmosphere is equivalent to 2.6 m of sea water, and therefore it can be seen that the thermal capacity of the ocean is equivalent to approximately 1600 atmospheres. However, during the seasonal cycle, only the upper 100 m of the ocean is affected significantly, and therefore the thermal capacity is approximately equivalent to only 38 atmospheric masses, giving a thermal relaxation time of approximately 3.5 years. This simple sum suggests that the ocean is responsible for a major modification of the seasonal cycle of temperature in the atmosphere. An example of the oceanic influence is shown in Figure 1.4. The flux of the long-wave radiation from the top of the atmosphere gives an indication of the seasonal change in temperature. In the Northern Hemisphere, where the largest ratio of continents to ocean (see Figure 1.9) exists (some 70% at 60°N), there is a large change of approximately 40 K in the radiation temperature during the seasonal cycle, whilst in the Southern Hemisphere between 30 and 60°S, where the continental area is less than 10%, there is only a small change in radiation temperature during the year. The most continental region benefits from an amelioration in its seasonal climate brought about by the oceans. Even in Siberia the winters are not as severe as they would be if the ocean did not exist!

1.5   Albedo

The planetary albedo is the ratio of the reflected (scattered) solar radiation to the incident solar radiation measured above the atmosphere. As this reflected radiation is lost immediately from the Earth–atmosphere system, the accurate determination of planetary albedo is important in the calculation of the amount of solar radiation absorbed by the system, and therefore the climate of the Earth. On average, the global albedo is 30% (±2%). This value has been determined from a number of satellite measurements over a period of a few years. The global albedo is not constant but varies on both seasonal and interannual time scales by approximately 2%. Recent observations have shown that the albedo is a maximum in January and a minimum in July. If the cloud and surface distributions were similar in both hemispheres, then no annual variation would be expected. However, because of the large seasonal cycle in snow extent over the Northern Hemisphere continents, particularly over the Eurasian land mass, and the seasonal cycle of cloud in the Northern Hemisphere, the global albedo has a seasonal component.

The latitudinal variation in albedo is determined by:

(i)   The elevation of the Sun.
(ii) The distribution and type of cloud.
(iii) The surface albedo.

At latitudes less than 30°, the planetary albedo is relatively constant at 25%, but it then increases with latitude to reach a maximum value of 70% over the poles. However, because the polar regions occupy less than 8% of the Earth's surface, the value of the global albedo is weighted to the albedo of the middle and low latitudes. It is noted that the latitudinal distributions in the albedo in the Southern and Northern Hemispheres are similar, despite the differences in land and sea distribution between the two hemispheres. This indicates that the cloud distributions in both hemispheres have a dominant influence over the surface variations in albedo in determining the planetary albedo (see Figure 1.5).

The long-term latitudinal variation in absorbed solar energy and the emitted planetary radiation is depicted in Figure 1.6a. The absorbed solar energy has maximum values of $300 \, \mathrm{W \, m^{-2}}$ in low latitudes. The planetary radiation, which is dominated by emission from the lower troposphere, shows a decrease with latitude which occurs at a slower rate than the decrease in the absorbed solar radiation. If the atmosphere were in radiative balance at each latitude, the two curves should be identical. However, the solar absorption exceeds the planetary emission between 40°N and 40°S, and therefore there is a net excess in low latitudes and a net deficit in high latitudes (i.e. poleward of 40°). This latitudinal imbalance in radiation implies that heat must be transferred from low to high latitudes by the circulation of the atmosphere and ocean. In effect, the atmospheric and oceanic circulations are maintaining higher atmospheric temperatures polewards of 40° than can be maintained by solar absorption.

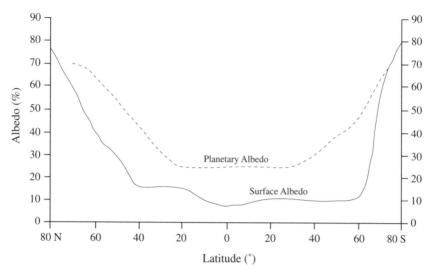

Figure 1.5 Planetary and surface albedo (%) expressed as a function of latitude. The planetary albedo is the fraction of incident solar energy reflected back into space, by the atmosphere and the Earth's surface

To obtain the transport of heat by atmosphere and ocean, the net radiation at each latitude, i.e. the difference between the absorbed solar radiation and the emitted planetary radiation, is integrated from pole to pole (Figure 1.6b). Thus, it is possible to determine the poleward transport of heat from satellite radiation measurements, but it is not possible to distinguish what proportion of the heat is transferred by the atmosphere and what proportion by the ocean. It is noted that the maximum transfer of heat occurs between 30° and 40° of latitude, and is approximately 5×10^{15} W.

Seasonal variations in the surface albedo are significant, particularly in the middle and high latitudes of the Northern Hemisphere, and arise from changes in the snow and cloud cover. At 60°N the albedo varies from 65% in February to less than 40% in July, as a result of the seasonal variation in snow cover. In the Southern Hemisphere the seasonal cycle is not so pronounced, but a minimum in albedo can be observed in March at 60°S, which is probably associated with the minimum sea ice extent around Antarctica. The emitted planetary radiation shows a strong asymmetry between the Southern and the

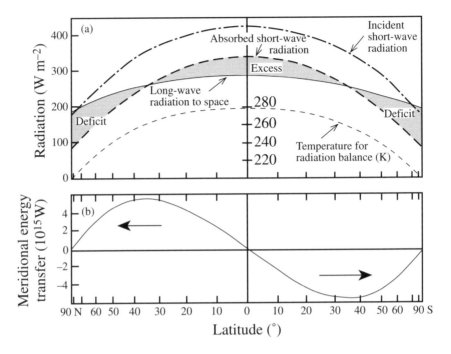

Figure 1.6 (a) Radiation balance of the Earth–atmosphere system as a function of latitude. The thin broken curve indicates an equivalent radiative temperature (centre scale) for each latitude band to be individually in balance with the absorbed solar radiation (heavy broken curve). The thin solid curve shows the actual observed long-wave radiation emitted to space as a function of latitude. (b) Poleward energy transfer within the Earth–atmosphere system needed to balance the resulting meridional distribution of net radiative gains (hatched) and losses (stippled) as indicated in (a). Note that the horizontal scale is compressed towards the poles so as to be proportional to the surface area of the Earth between latitude circles

Northern Hemispheres, which is the result of the different land–sea distribution in each hemisphere.

The net radiation budget, illustrated in Figure 1.7, indicates substantial seasonal variation as the result of the seasonal solar distribution. The zero contour varies from 15 to 70º latitude, thus showing that the zone of maximum heat transport will also vary during the seasonal cycle. The maximum net radiation gain varies between 30ºS and 30ºN, and is larger in the Southern Hemisphere than in the Northern Hemisphere.

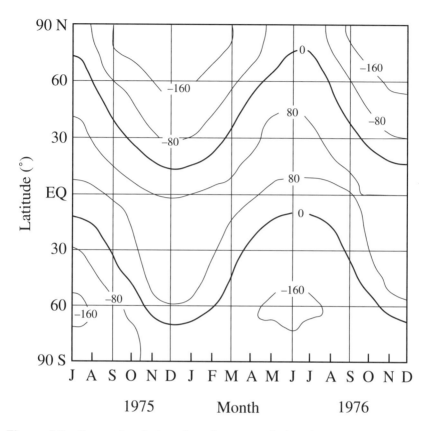

Figure 1.7 Seasonal variation of zonal mean net (solar − long-wave) radiation budget, $W\,m^{-2}$, measured by satellite from July 1975 to December 1976

The surface albedo, as has been shown, makes an important contribution to the planetary albedo, although the cloud distribution is, without doubt, the dominant influence. However, the amount of solar radiation absorbed by the surface, and therefore the surface temperature, are strongly influenced by the albedo of that surface. The albedo of any particular surface depends on the following factors:

(i)   The type of surface.
(ii) The solar elevation and the geometry of the surface relative to the Sun.
(iii) The spectral distribution of the solar radiation and the spectral reflection.

Table 1.3 shows typical values of the surface albedo. It is noted that,

Table 1.3 Albedos for different terrestrial surfaces

Type of surface	Albedo (%)
Ocean	2–10
Forest	6–18
Grass	7–25
Soil	10–20
Desert (sand)	35–45
Ice	20–70
Snow (fresh)	70–80

for a calm sea surface, the albedo is only 2%, and therefore the oceans are generally very good absorbers of solar radiation in comparison with most types of land surface. Forests and wet surfaces generally have low albedos, whilst deserts and snow-covered surfaces have high albedos. The surface albedo is also a function of the spectral reflectivity of the surface, and variations between visible and near infra-red can be substantial. For instance, though freshly fallen snow has an albedo greater than 90% in the visible, at 1.5 μm in the near infra-red, its albedo is only 25%. Vegetation usually has a low albedo in the visible region, but a high albedo in the near infra-red. Although it is useful to be able to derive a single albedo for the solar spectrum, it must be remembered that the solar spectrum incident on the surface changes with solar elevation and cloud cover, and that the albedo will also change.

1.6 The topography of the Earth's surface

The Earth is a spheroid which is slightly distorted by rotation. This distortion amounts to approximately 0.3% of the Earth's mean radius, with the equatorial radius being greater than the polar radius. Because of the irregularities in the Earth's surface, geophysicists have defined a surface known as the geoid. The geoid is a surface of equipotential gravity. The height of this surface varies with position relative to the centre of mass of the Earth, but *along* the surface there is no gravitational force, i.e. the direction of the gravity force is everywhere perpendicular to the equipotential surface.

The ocean surface would be an equipotential surface if winds, tides, currents and density variations did not produce deformations in its surface. Oceanographers, therefore, have defined their reference level in terms of the mean sea level, after the removal of tides and meteorological influences. The mean sea level has been obtained by long-term observations of the sea level at coastal stations around the world. With reference to the mean sea level, the continental surface has a mean altitude of 840 m, whilst the mean depth of the ocean is 3800 m, which implies that the average height of the Earth's crust is 2440 m below sea level.

The relative topography of the Earth's surface is depicted in Figure 1.8, where it is noted that the frequency distribution has two maxima; one corresponding to the mean height of the continents and the second corresponding to an ocean depth of approximately 4500 m. This diagram shows that the most frequent height of the Earth's crust corresponds to the ocean bottom. The ocean occupies 70.8% of the Earth's surface, and 76% of this ocean area has depths between 3000

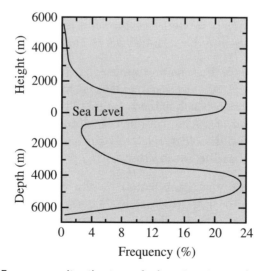

Figure 1.8 Frequency distribution of elevation intervals of the Earth's surface

and 6000 m. These depths are known as the abyssal depths of the ocean. Of the remaining 24% of the ocean area, 7.5% is occupied by the continental shelf, which is roughly confined to depths of less than 200 m, and 15.3% is occupied by the continental slope, which is the transition region between the continental shelf and the abyssal ocean. Only 1.1% of the ocean has depths greater than 6000 m and only 0.1% has depths greater than 7000 m. This small percentage is associated with ocean trenches, whose depths plunge down to 11 500 m in the Mindanao Trench.

The relative distributions of land and ocean as a function of latitude are shown in Figure 1.9. It is again noted that the land/sea distribution is asymmetrical, with over 80% of the total global land area being located in the Northern Hemisphere and 63% of the global ocean area in the Southern Hemisphere. The continental land masses of North America and Eurasia contribute to the dominance of land over ocean between 45 and 70°N. At corresponding latitudes in the Southern Hemisphere are regions where land occupies less than 3% of the total area, with essentially no land between 55 and 60°S. Poleward of 70° latitude the situation is reversed, with the Northern Hemisphere being occupied by the Arctic Ocean and the Southern Hemisphere being occupied by the continent of Antarctica. It is believed that the presence of a continent in a polar region is a significant factor in causing glacial cycles over the past 60 million years. An interesting property of the land/ocean distribution is that 95% of all land points have antipodes in the ocean.

The distribution of the major topographic features of the globe is shown in Figure 1.10. The most extensive feature is the continental slope, or escarpment, which extends for more than 300 000 km, with an average slope of 4°. The continental slope marks the geological boundary between the older, relatively thick, granitic continental crust and the young, basaltic ocean crust.

The second most important feature is the mid-ocean ridge system which has a length of approximately 75 000 km. The typical width of the ridge is about 500 km and the height ranges from 1 to 3 km above the abyssal depths. The ridges are regions of newly formed ocean crust, with the crust moving outwards from the central part of the ridge system. A rift valley approximately 20–50 km wide, along the central part of the system, is the region of major crustal formation, and it is

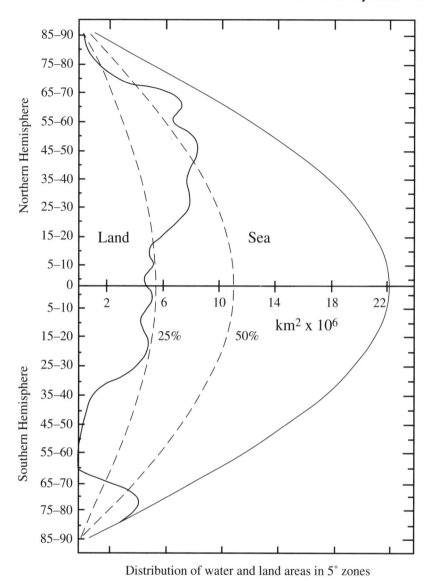

Distribution of water and land areas in 5° zones

Figure 1.9 Distribution of water and land areas in 5° zones for the Northern and Southern Hemispheres

frequently associated with volcanic and seismic activity. The ocean ridges are of importance in the movement of deep-water masses because they split the oceans into smaller ocean basins. In the Atlantic Ocean the eastern abyssal basins are separated from the western

Figure 1.10 Major topographic features of the ocean and the continents

Land above 2000 metres

Continental Shelf (200 metres)

200-4000 metres

> 4000 metres

Major trench

abyssal basin by the mid-Atlantic ocean ridge. Connections between the basins, however, occur in the fracture zones which are orientated perpendicular to the axis of the ridge. Two important fracture zones in the Atlantic Ocean are the Romanche gap close to the equator, which allows Antarctic Bottom Water to flow into the eastern basins, and the Gibbs Fracture zone at about 53°N, which allows the North Atlantic Deep Water to flow westward into the Labrador basin. Furthermore, the ridge system has an influence on oceanic flows above the level of the ridge topography and in some regions, particularly in the Southern Ocean, its effect extends to the ocean surface currents.

The relatively low relief of the land masses has already been noted and only 11% of the land area is above 2000 m. However, there are two significant features associated with the Tertiary mountain building which are worthy of further comment. These are the Western Cordillera of the American continent, which has an altitude of over 6 km above sea level, and the Himalaya mountain chain, with altitudes in excess of 8 km. These mountain ranges are very important in their influence on the general circulation of the atmosphere. Although the atmosphere extends well beyond the highest peaks, 50% of the atmospheric mass occurs below 5 km, and therefore these large mountain chains have a major disruptive effect on the flow of the lower atmosphere. The Western Cordillera has a predominantly north–south orientation and this disrupts the westerly flow for over 5000 km downstream. The Himalayas have an east–west orientation which results in an intensification of the zonal flow and an inhibition of the northward penetration of the Asian monsoon. Not only is the flow directly affected by the mountain barriers but, because most of the water vapour in the atmosphere is confined to the lowest 5 km, the geographical distribution of the hydrological cycle is also significantly affected. Furthermore, the release of latent heat by condensation into the atmosphere provides a major source of energy for the general circulation of the atmosphere, and the mountain chains will influence the location of heat sources in the atmosphere and the associated low-level atmospheric flow.

A feature worthy of note is the comparison of the horizontal scales of the ocean and atmosphere with the vertical scales. The ocean basins have horizontal dimensions of 5000 km in the Atlantic Ocean to over 15 000 km in the Pacific Ocean, whilst the average depths are approximately 3.8 km. Thus, the vertical scale is small compared with the horizontal scale and the ratio between them, known as the aspect ratio, is at most 1 part in 1000 and, more typically, 1 part in 10 000. The atmosphere has 99% of its mass between the sea surface and a height of 30 km, and compared with the circumference of the globe, the aspect ratio is again less than 1 in 1000. Thus, both atmosphere and ocean are indeed very thin when compared with their horizontal dimensions.

It has been shown that the topography of the land masses has a significant effect on the circulation of the lower atmosphere. In a different way, the shape and size of the ocean basins have profound effects on the circulation of the water masses. Except for the region lying between 50 and 60°S, the ocean is bounded by continental masses, and therefore zonal (i.e. west–east) flows are restricted to the Southern Ocean. The presence of continental margins produces intense meridional, i.e. north–south, flows in the western regions of the basins, as evidenced by the Gulf Stream in the North Atlantic and the Kuroshio current in the North Pacific. All the ocean basins are connected via the circumpolar flow of the Southern Ocean. In contrast, the flow into the Arctic Ocean from the Pacific Ocean is inhibited by the shallowness of the Bering Sea, whilst from the Atlantic there are 'sills' which control the flow between Greenland and Iceland, and between Iceland and Scotland.

There is a deep-water connection between the Indian Ocean and the Pacific Ocean, through the Indonesian Archipelago, which has important influences on the flow around the Australian continent, and on the exchange of surface and intermediate water masses between the two ocean basins.

The width of the ocean basins has a marked effect on the ocean–atmosphere interaction. The relatively narrow North Atlantic Ocean is exposed to drier air from the surrounding continents than the Pacific Ocean, and therefore evaporation is approximately twice as large in the North Atlantic Ocean as in the Pacific Ocean. This high evaporation rate is a factor in a larger heat loss from the North Atlantic Ocean than the Pacific Ocean, and hence dense, cold water masses are formed in the North Atlantic Ocean rather than in the Pacific Ocean. Thus, the deep-water vertical circulation in the North Atlantic Ocean is much more intense than the sluggish vertical circulation observed in the Pacific Ocean. It will be shown in Chapter 5 that this has consequences for the heat transport in the ocean. In the context of evaporation, the marginal seas, such as the Mediterranean Sea and the Red Sea, are important regions for the formation of dense water masses because of their high evaporation rates.

The continental shelf, though occupying a relatively small area of the ocean, is of considerable importance both to man and to life in the ocean. The intense flows, associated with winds and tides in the continental shelf region, result in a mixing of the nutrients from the continental shelf sediment into the water column. These nutrients, in conjunction with the ready availability of oxygen and sunlight in these shallow sea areas, produce an abundance of phytoplankton. These regions of high primary production are very important in the maintenance of marine food chains and contain the most intensive fisheries in the world. The sedimentation rates on the continental shelves are also high, being of the order of 1 m per 1000 years in comparison with 0.1–10 mm per 1000 years in the deep ocean. Much of the sediment deposited is derived from river-borne material and this contains large quantities of phosphorus, which is an element necessary in the maintenance of productivity in the ocean.

In conclusion, it has been shown that the surface topography, and the shape and interlinkage of the ocean basins, have a significant effect on the circulation of both the atmosphere and the ocean which, in turn, affect the biological productivity of the land and the ocean. In the geological past, different ocean geometry and surface topography would have been partly responsible for the many varied climatic patterns which have occurred since the formation of the Earth.

2

Composition and Physical Properties of the Ocean and Atmosphere

2.1 Evolution of the atmosphere and ocean

Direct evidence of the form and composition of the early Earth have long since been destroyed and clues can only be gleaned from an understanding of geochemical processes. Various radioactive elements can be used to evaluate the timing of development of the atmosphere, and understanding of the behaviour of certain elements allows estimation of the atmospheric and oceanic composition through time. The Earth and other planets were formed some 4.6×10^9 years ago (4.6 Ga) via an accretionary process involving collisions between cosmic dust through to bombardment of planetesimals by meteorites, eventually resulting in the solar system that we know today. This accretion is inferred to have continued for some 150 million years and the extensive bombardment of the young Earth continued until at least 3.8 Ga before present.

The accepted paradigm for oceanic evolution involves precipitation from a transitory steam atmosphere created by the continued impact and heating of the proto-Earth surface during planetary accretion. The ocean precipitated out as the Earth cooled and has remained liquid throughout intervening geological time. This has allowed development of life on Earth and puts lower limits on any climatic variations inferred to have taken place, since the oceans have never frozen completely. Upper constraints to surface temperatures and atmospheric structure are imposed by the presence of the large hydrosphere on Earth. The present-day atmospheric temperature structure means that water vapour precipitates in the upper troposphere and the concentration of water vapour in the stratosphere is very low. This retains water within the lower atmosphere and limits passage of H_2O to the upper atmosphere, where extensive photodissociation would occur and the H_2 evolved would escape to space due to its low atomic mass.

Comparison of the rare, inert gas composition of the atmosphere with that of the Sun and average solar system abundances demonstrates that the Earth lost its primordial atmosphere during the accretionary process, which involved heating and melting of the new planetary mass. The current atmosphere formed as a consequence of degassing of the cooling planet Earth once accretion had ceased. Again, rare gas composition can be used to determine the timing of the atmospheric development; 80–85% of the present-day rare gas content degassed from the Earth's interior in the first million years of history, the remainder has leaked constantly out during the intervening eons.

Volcanism is the major route for degassing of volatile material from the inner Earth. Present-day volcanic activity produces H_2O, CO_2, SO_2, N_2, H_2 and Cl_2 in substantial quantities. The composition of early volcanic gases is a matter of debate and there may have been a substantial component of NH_3 and CH_4 depending on the extent of oxidation of the source rocks in the upper mantle.

The early solar system was illuminated by a weak, young Sun that only delivered $\sim$75% of the present-day energy to the Earth's surface. To provide a non-frozen ocean there must have been an increased 'greenhouse' effect provided by CO_2. One-dimensional climate models require CO_2 levels of 500 times present atmospheric levels (PAL) at 2.75 Ga. However, recent estimates of atmospheric CO_2 levels from study of ancient soils (palaeosols) suggest upper limits on Archaean partial pressure of carbon dioxide of 100 PAL. The additional planetary warming must be derived from other greenhouse gases such as NH_3 and CH_4 or albedo variations. The same palaeosol samples allow estimation of the early atmospheric O_2 content as being approximately 1.4×10^{-4} PAL. This O_2 was mainly derived from the photodissociation of H_2O in the upper atmosphere.

The behaviour of CO_2 is crucial to our understanding of early Earth history. Early levels were elevated and, by the Cambrian era ($\sim$600 Ma), CO_2 levels were close to the PAL. The main reason for the extraction of CO_2 out of the atmosphere was the development of life forms in the ocean that sequestered carbon into organic and later inorganic (calcium carbonate) forms and buried it in sedimentary formations on the sea floor. Life must have originated as the planetary bombardment slowed after 3.8 Ga; the oldest known fossils are from blue-green algae dated at 3.5 Ga. The precursors to these organisms were almost certainly high-temperature ($>70^\circ C$) chemosynthetic archaea living in specialized niches, such as submarine hydrothermal vent systems in areas of underwater volcanism.

Photosynthesis involves the oxidation of liquid water and the reduction of carbon dioxide to carbohydrate and release of O_2: $6CO_2 + 6H_2O \leftrightarrow (CH_2O)_6 + 6O_2$. The reverse of this process is respiration which is carried out by all life forms. There is extensive evidence from palaeosols and Fe-rich sedimentary formations that atmospheric O_2 levels did not achieve appreciable levels until 2 Ga and approached present-day levels by 1.5 Ga. Until then, all O_2 produced was consumed in the ocean by respiration and reduction by dissolved chemical species such as Fe (II). Once O_2 started to accumulate in the atmosphere, the production of ozone (O_3) could occur. Ozone plays a vital role in the upper atmosphere by shielding the Earth's surface from damaging ultraviolet (UV) radiation, allowing life to move from the purely oceanic realm and begin to colonize the land masses. The evolution of land plants and the terrestrial biosphere would extract more CO_2 from the atmosphere and provide an additional source of O_2.

The present-day atmosphere is dominated by chemical species that demonstrate the existence of life on Earth. In the modern system,

Table 2.1 The carbon budget for the Earth

Reservoir	Mass of carbon (10^{14} kg)
Atmosphere	7
Ocean	390
Land	
Vegetation	6
Soil	15
Sediments	200 000
Fossil fuels	50

99.8% of the present-day carbon is locked up in sedimentary rocks of mainly biogenic origin, whereas only 3.5×10^{-3}% of the carbon is in the atmosphere as CO_2 (Table 2.1). The atmospheric O_2 levels of 21% represent the standing crop of photosynthetic products maintained by global primary production. This contrasts starkly with the early Earth atmosphere.

Comparison of Earth with its nearest neighbours in the solar system is illuminating. The starting products for Venus, Earth and Mars were likely to be similar as all three planets were formed by the same accretionary process. Venus has not developed the atmospheric temperature inversion, between troposphere and stratosphere, essential for maintaining liquid H_2O. Venus has a dense CO_2-rich atmosphere with a small component of water vapour and extremely high surface temperatures of 460ºC. Studies of the heavy isotope of hydrogen, deuterium, suggest that a large proportion of the early Venus water budget was lost to space via photodissociation in the upper atmosphere. In contrast, Mars has a very thin CO_2-dominated atmosphere and cold surface temperatures, with solid CO_2 ice caps. There is some evidence from fluvial-type deposits that running water was in existence on the surface of Mars at 3.7 Ga, but has been absent since this time. Again, photodissociation in the upper atmosphere and escape of H_2 to space is invoked as a mechanism for water loss. The O_2 evolved by this process has oxidized the surface rocks to their characteristic red colour.

As already intimated, the atmospheric composition of greenhouse gases plays a major role in the Earth's climate. Evidence from fossils deposited on the sea floor allows estimation of sea-water temperatures through geological time. The mean surface temperature has decreased throughout the first few billion years of Earth history. The first geological evidence for glaciation occurred between 2.5 and 2.3 Ga. However, earlier glaciations cannot be ruled out due to the paucity of geological evidence for this period. Climate variations have occurred throughout geological time and the geological record becomes more and more complete as the modern era is approached.

The Earth has undergone a major cooling episode over the last 65 Ma. This is associated with a decrease in CO_2 levels and the development of permanent ice sheets in the polar regions. Approximately 2 Ma ago the Earth went into a series of glacial–interglacial cycles, the most recent glaciation ending about 10 000 years ago. The pace-maker behind these climate oscillations has been attributed to variations in orbital parameters that determine surface insolation named Milankovitch cycles, discussed in Chapter 1.

Whatever drives the timing of glacial–interglacial cycles, it is apparent from evidence recorded in ice cores, deep-sea sediments, corals, varved lacustrine sediments and climate modelling that there have been changes in atmospheric CO_2 levels over the last 1 Ma. Atmospheric CO_2 levels were as low as 0.7 PAL during the last glacial maximum. Much effort has gone into understanding atmospheric and oceanic variability over the last few hundred thousand years in an attempt to understand the short-term climate variability. This may allow prediction of the consequences of fossil fuel burning and deforestation that act to increase atmospheric CO_2 levels by 0.4% per annum.

The development of the ocean

Once accretion of the Earth stopped, the surface geological process of crustal generation and plate tectonics that we are familiar with today could begin. Zircons are resistant mineral grains from continental crust that can withstand intense weathering and metamorphic processes. The oldest zircons from Western Australia have been dated at 4.1 to 4.3 Ga, giving a limit to the onset of continental crust generation. In contrast, the oldest oceanic crust is of the order of 200 Ma in age, a mere 0.4% of Earth's history. New oceanic crust is continually created at spreading centres located along the mid-ocean ridges, and is destroyed at the convergent margins where the ocean crust is subducted into the Earth's mantle. The rate of spreading is typically a few centimetres each year, with the fastest rates at the East Pacific Rise and the slowest at the mid-Atlantic ridge.

The amount of water in the hydrosphere is in steady-state balance between the input of juvenile, magma-derived water from creation of new crust and the loss by photodissociation in the upper atmosphere. Sea level has varied through time by several hundred metres as various amounts of water are locked up in the ice caps. In general, the rule of isostatic equilibrium means that the oceans fill the oceanic basins and flooding of continental crust is not extensive.

2.2   Present-day composition of sea water

Dissolved salts comprise about 3.5% by volume of sea water. The salts are present because of the relative ease with which ionic compounds dissolve in water. Originally, it was believed that the salts dissolved in the ocean were derived from the weathering of continental rocks, the products of which were then transported by rivers to the ocean. Robert Boyle made measurements of the composition of river water and he showed that salts could be detected. However, with more accurate methods of analysis, it was found that, although the river waters contained potassium, magnesium, calcium and sodium ions, they were deficient in dissolved carbon dioxide (i.e. carbonate and bicarbonate), chlorides, bromides and iodides. In fact, if the ocean salts had been derived from this one source, their composition would have to be totally different from that observed (Table 2.2). The ions missing from land weathering products originated from volcanic and hot spring discharges, and atmospheric sources.

The observed distribution is not, however, simply determined by the inputs from weathered rock and volcanic emissions, because ions are simultaneously being lost from the water column by sedimentation on the ocean floor and continental shelves, and by transfer to the atmosphere and biosphere. The present-day composition is, therefore, a geochemical balance between the sources and sinks of the ions. It is believed that the ocean has been in equilibrium for 100×10^6 years and that the total salt content of the ocean has not changed

Table 2.2   Composition of standard ocean water (salinity = 34.7 psu)

Constituent ion	Symbol	Mass of ion in grams for 1 kg of sea water	Percentage of total salt
Chloride	Cl^-	19.215	54.96
Sodium	Na^+	10.685	30.58
Sulphate	SO_4^{2-}	2.693	7.70
Magnesium	Mg^{2+}	1.287	3.69
Calcium	Ca^{2+}	0.410	1.17
Potassium	K^+	0.396	1.13
Bicarbonate	HCO_3^-	0.142	0.41
Bromide	Br^-	0.067	0.19
Boric acid	H_3BO_3	0.026	0.07

significantly in that time. This is based on the fact that the pH of the ocean must have remained reasonably constant to enable the present life forms to evolve. It is noted that a 1% decrease in the sodium ion concentration would result in a strongly acidic ocean with a pH of 2.6. However, the ratios of the constituent salts in the ocean have shown considerable fluctuations. For example, the Na^+/K^+ ratio was approximately 1 in the pre-Cambrian period ($\sim 10^9$ years ago) and now it is 28. The changes in the chemical composition are related to the evolution of organisms which extracted materials such as calcium and silicon for shell building, and so altered the steady-state balance.

The first comprehensive analysis of the composition of sea water was made by William Dittmar who analysed 77 samples of sea water, collected by HMS *Challenger* between 1872 and 1877. The samples were representative of all the major ocean basins and were taken at depths ranging from the surface to the abyssal layers. Dittmar analysed the samples for the eight major constituents (see Table 2.2) and showed that, within narrow limits, the ratios of these major constituents were in constant proportion. Together with boric acid, these eight constituents comprise 99.9% of the total dissolved salts in the ocean. The remaining 0.1% represents virtually all the stable elements of the periodic table and they do not generally obey the rule of constant proportions. The most prominent of the minor constituents are strontium, silicon, iron, lithium, phosphorus and iodine, as well as the dissolved gases oxygen and nitrogen, all of which are fundamental to the chemical and biological processes in the ocean.

Dittmar's experimental proof that the major constituents were present in constant proportions to each other throughout the world's oceans, including the Mediterranean and Baltic Seas, allowed oceanographers to define the salt content of the ocean as a single physical quantity, known as salinity. From a determination of the salinity and the rule of constant proportions, the concentration of any one of the major constituents can be determined. Furthermore, in order to obtain a value of the salinity of sea water, it is only necessary to measure one of the major constituents. The most easily measured constituent is the chloride content of sea water. In practice, the chlorinity (Cl) is determined, which is the amount of chlorine (grams) in 1 kg of sea water, where both bromine and iodine are replaced by chlorine. The relationship between salinity and chlorinity is given by

$$\text{salinity } (\%_0) = 1.80655 \, Cl$$

By titration against silver nitrate, chlorinity can be measured and the salinity obtained to an accuracy of ± 0.01.

Since the 1980s, the titration method has been replaced by the direct measurement of the electrical conductivity of sea water. The definition of salinity has also been redefined to be the ratio of the electrical conductivity of sea water with a standard solution of potassium chloride at standard pressure and temperature. This ratio is expressed in Practical Salinity Units (psu), where 1 psu $= 1.00510 \times$ salinity. The measurement accuracy of 0.001 is a considerable improvement over the titration method.

2.3   Introduction to gases and liquids

The relationship between pressure, p, temperature, T, and the density, ρ, for an ideal gas is:

$$p = \rho R T \tag{2.1}$$

where R is the gas constant for the constituent gas having molecular weight m. $R = R^*/m$, where R^* is the universal gas constant.

The derivation of the ideal gas equation depends on the assumption that the molecules in an ideal gas are so far apart that intermolecular forces can be disregarded except during the very brief collisions. The strong intermolecular forces between ideal gas molecules allow them to be treated as hard spheres and their collisions may be considered to be inelastic. Many gases approximate to ideal gases under certain temperature and pressure conditions.

Such simple assumptions are invalid when dealing with the liquid and the solid state. In a liquid the intermolecular forces cannot be neglected or generalized, since they vary from one liquid to another. It is not, therefore, possible to produce an ideal equation of state for an ideal liquid. The long-range attractive intermolecular forces in a liquid allow a much more efficient packing of the molecules than is found in a gas. Hence, liquids have densities approximately 100–1000 times greater than the density of a gas. For instance, air at normal atmospheric pressure and 0°C has a density of $1.28 \, \text{kg m}^{-3}$, whilst sea water has a density of $1028 \, \text{kg m}^{-3}$.

At very short ranges, of the order of 0.1 nm, the intermolecular repulsive forces become dominant and there is, therefore, a limit to the close packing of molecules. Solids may have densities approximately 15% greater than liquids, but this increase is achieved by more efficient, long-range packing arrangements rather than a smaller intermolecular separation. In this respect, the densities of water and ice are anomalous. Ice has a density which is 9% less than liquid water, and thus ice can float on water.

The liquid state has properties similar to the solid state on a short time scale in that small clusters of molecules are ordered in a similar way to the solid-state lattice. However, on a large time scale the liquid resembles a gas in that it is highly disordered. Liquids may therefore be considered, in some respects, as being an intermediate state between the extreme ordering of the solid-state lattice and the randomness of the gaseous state. On this longer time scale, a liquid and gas will not support a shear stress and, as a result, they will yield and flow. In this case we may call the gas or liquid a fluid and we can then describe mathematically properties of motion without referring to the substance as liquid or gas.

Gases, furthermore, have the property that they will expand to fill the space available, and therefore it is impossible for a gas to have a free surface. In this context it is not meaningful to refer to the height or thickness of the atmosphere. A liquid, on the other hand, always has a free surface, and therefore it is possible to define the interface between atmosphere and ocean. Another important property of a gas is that it is readily compressible due to the large distances between molecules. However, because of the close packing in a liquid, it is difficult to compress a liquid without resorting to large pressures.

The equation of state for the atmosphere

The atmosphere can be described as a mixture of gases, predominantly oxygen, nitrogen, carbon dioxide, argon and water vapour (Table 2.3). Of these major constituents, only water vapour is a significantly variable constituent, and it is therefore considered separately to the other gases. The equation of state for dry air can be obtained by summing the partial pressures of each constituent gas. If each gas occupies a similar volume, V, and has a mass of M_n, then the partial pressure P_n is given by

Table 2.3 Composition and scale heights of terrestrial atmospheres

	Major constituents (% of molecules)	Mean molecular weight (amu)[a] (equation 2.2)	Gas constant (J kg^{-1} K^{-1})	Scale height (km) (equation 2.4)
Venus	95.5% CO_2 4% N_2	43.35	192	14.9[b]
Earth	78% N_2 21% O_2 96% CO_2	28.97	287	7.4
Mars	2.5% N_2 1.5% Ar	43.55	191	10.9

[a]amu = atomic mass units.   [b]Lower atmosphere.

$$P_n = \left(\frac{R^*}{m_n}\right)\frac{M_n}{V}T$$

where m_n is the molecular weight of the gas.

The total pressure for N constituent gases is

$$p = \sum_{n=1}^{N} P_n = \frac{R * T}{V} \sum_{n=1}^{N} \frac{M_n}{m_n}$$

The mean density of the mixture, $\bar{\rho} = \sum_{n=1}^{N}(M_n/V)$. Therefore, $p = \bar{\rho}(R^*/\bar{m})T$, where

$$\frac{1}{\bar{m}} = \frac{\displaystyle\sum_{n=1}^{N}(M_n/m_m)}{\displaystyle\sum_{n=1}^{N}M_n} \tag{2.2}$$

or $p = \rho R_d T$, where R_d is the gas constant for dry air. R_d has a value of $287\,\text{J K}^{-1}\,\text{kg}^{-1}$.

Comparing equation 2.2 with 2.1, it is noted that the equation of state for a mixture of gases is similar to a single constituent equation of state, provided that the molecular weight is replaced by the mean molecular weight for the mixture.

Using an example for the Earth with the following values:

Constituent	M_n	m_n
N_2	75.51	28.016
O_2	23.14	32.00
Ar	1.28	39.24
CO_2	0.02	44.01

For the major constituents:

$$1/\bar{m} = \frac{75.51/28.016 + 23.14/32.00 + 1.28/39.24 + 0.02/44.01}{99.95}$$

hence $\bar{m} = 28.96$

The equation of state for sea water

The equation of state for liquid water is a function of three parameters: temperature, pressure and density. For sea water, the relationship involves the properties of all the constituent salts and is therefore extremely complicated. However, the observation of the constancy of composition for the major elements of sea water allows the individual constituents to be replaced by one variable, the salinity. As discussed earlier, there is no general statistical theory for all liquids, and therefore the relationship has been determined by laboratory measurements and the results are tabulated. The *in situ* density of sea water can then be given as $\rho = \rho(T, P, S)$. Table 2.4 shows the values of *in situ* density for extreme conditions in the ocean. It is noted that *in situ* density increases both with increasing pressure and salinity, but decreases with temperature. However, the variations in density are less than 7%, and therefore it is common practice to define

$$\sigma(T, P, S) = \rho(T, P, S) - 1000 \text{ kg } m^{-3}$$

Table 2.4   Density of ocean sea water for extreme ranges of temperature and pressure in the ocean

Depth (m)	Pressure (Pa)	Density (kg m^{-3})	
		$T = 273$ K, $S = 35$ psu	$T = 303$ K, $S = 35$ psu
0	10^5	1028	1022
10 000	10^8	1071	1060

The values of $\sigma(T, P, S)$ can be obtained from standard tables (e.g. UNESCO, 1981) or from polynomial relationships. However, for many purposes the effect of pressure is eliminated by using a reference pressure, i.e. mean atmospheric pressure, to define density. This quantity is referred to as sigma-t or σ_t.

Figure 2.1 shows the relationship of σ_t as a function of temperature and salinity at atmospheric pressure. Although in all cases density increases with increasing salinity and decreasing temperature, the relationship is non-linear. At temperatures of 25–30°C, typical of the equatorial ocean, the isopycnals (lines of constant density) are reasonably linear and a simple relationship for σ_t can be used. However, at low temperatures a given salinity change produces a larger change of

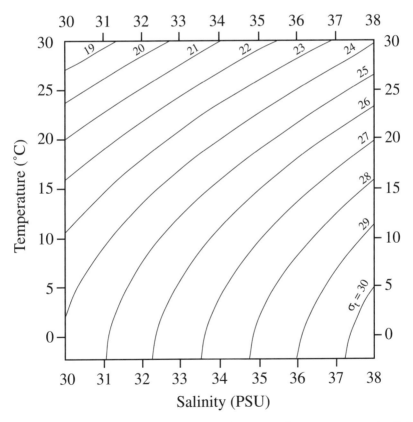

Figure 2.1   Density of sea water (σ_t) as a function of temperature and salinity

density than the corresponding change in salinity at high temperature. This physical property of the equation of state results in a greater sensitivity of density to salinity changes at high latitudes and is an important influence in the formation of bottom water. Another interesting property of the equation of state is that two water masses, with the same density but having a different temperature and salinity (see Figure 5.14), when mixed will produce a denser water mass than the original water masses. Again, this is an important process in the formation of dense bottom waters in polar regions of the ocean. If the ocean was well mixed in temperature and salinity, then the non-linearity of density would result in the increased density of the world's oceans and an inevitable decrease in sea level.

Figure 2.2 Changes in the maximum density temperature as a function of salinity (solid line). Changes in the freezing point as a function of salinity (dashed line)

The anomalous behaviour of pure water compared with other liquids has already been briefly mentioned. In particular, the maximum density of liquid water at 4°C and the expansion of water on freezing are two of its remarkable properties. Other properties include the high dielectric constant, which results from the high electrical dipole with a relatively small molecular volume. Because the interionic force of attraction varies inversely with the dielectric constant, the attraction between ions is diminished in water, and hence ionic components are easily dissolved in water. The presence of the dissolved salts in sea water increases the electrical conductivity by 10^6 over that of pure water. This property has been used to advantage in the accurate electrical measurement of salinity in sea water, which is now the most reliable technique available to oceanographers. Additional remarkable features of water are its high specific heat, its high boiling point compared with other hydrogen-based liquids (e.g. 213 K for H_2S and 240 K for NH_3) and its large latent heat of vaporization of 2.3 $\times 10^6$ J kg^{-1}, compared with 0.55×10^6 J kg^{-1} for H_2S and 1.4 $\times 10^6$ J kg^{-1} for NH_3. This property allows water to be a major factor in the transfer of energy between the ocean and atmosphere.

The influence of salinity, on the properties of water, will be shown to be very important. Figure 2.2 shows the temperature of the maximum density of sea water and the freezing point as a function of salinity. For salinities less than 24.69 psu, water reaches a maximum density before the freezing point is reached, and therefore, between the temperature of maximum density and the freezing point, there is an anomalous decrease in density with decreasing temperature. However, for a water mass having a salinity in excess of 24.69 psu, which is true of all ocean water, the freezing point is reached before the maximum density of sea water. In this case there is no anomalous behaviour of sea water, and the densest water masses (for a given salinity) are associated with the lowest surface temperatures. For most ocean water masses, this temperature ~ -2°C. A second property of salinity is the effect on the specific heat of water. For pure water, at 0°C, the specific heat is 4218 J kg^{-1} K^{-1}, whilst for sea water it is 3940 J kg^{-1} K^{-1} for a salinity of 35 psu, and therefore sea water is 6% less efficient in storing heat than an equivalent mass of fresh water.

The propagation of sound in sea water is given by $C = \sqrt{K/\rho}$, where ρ is the density and K is the compressibility of sea water. They are both dependent on the temperature, pressure and, to a lesser extent, salinity.

$c(z)$ m s^{-1}

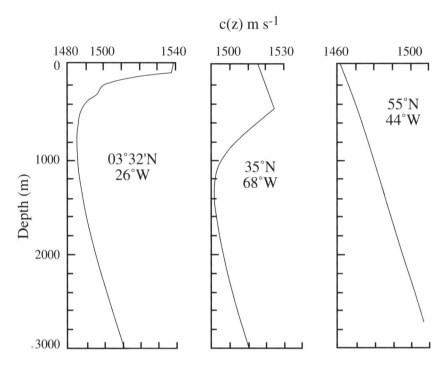

Figure 2.3 Velocity of sound as a function of depth, for tropical (left), temperate (centre) and sub-polar (right) oceans. Between the surface and 1000 m, temperature is the dominant effect, whilst pressure becomes dominant at depths greater than 1000 m

Figure 2.3 shows the variation in sound velocity for typical temperature and salinity profiles in the ocean. In the deep ocean (>1500 m), because of the uniformity of temperature and salinity (Figure 2.4), the sound velocity is dominated by pressure and therefore increases with depth. In the upper ocean where large variations of temperature and, to a lesser extent, salinity occur, the sound velocity displays more variability. The large decrease of temperature with depth, especially through the thermocline, causes a decrease in velocity with depth which is larger than the opposing pressure effect. Hence, a minimum velocity is observed at a depth of approximately 1 km in profiles where the thermocline is present. This minimum is very important for the propagation of sound and is known as the SOFAR channel. Sound waves emitted either upwards or downwards from this level will exhibit refraction into the sound channel because of the variation of the velocity of sound with depth. In this region, therefore, sound energy is

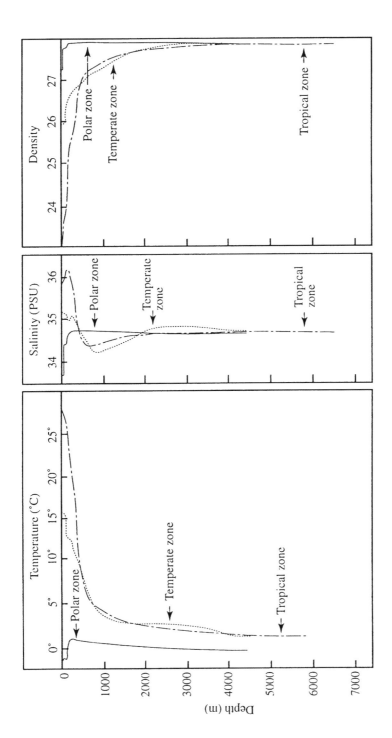

Figure 2.4 Temperature, salinity and density (σ_t) as a function of depth in the tropical, temperate and polar oceans

trapped in a wave guide, and a signal may propagate over hundreds and even a thousand kilometres before being significantly attenuated. This property has been used by oceanographers for the distant monitoring of SOFAR floats from research ships and shore bases (see page 000 and Figure 6.2). Experiments are now under way to determine the temperature of the global ocean by monitoring the travel times of low-frequency sound pulses from the Southern Ocean to the east and west coasts of North America.

2.4   Hydrostatic equilibrium

An observer descending from the outer reaches of the atmosphere to the Earth's surface with a barometer would note an increase in pressure. This increase in pressure would not be linear but would show an exponential increase with depth. Though pressure itself is a scalar quantity, the vertical variation of pressure in the atmosphere produces a pressure force which acts in the direction of decreasing pressure (i.e. upwards). This observed pressure-gradient force would be able to accelerate a parcel of air upwards at $\sim 10 \, \mathrm{m \, s^{-2}}$. In the absence of other forces, this acceleration would produce a velocity of $\sim 5000 \, \mathrm{km \, hour^{-1}}$ at a height of 100 km. Though the atmosphere is always in a state of motion, measured vertical velocities are rarely larger than $10 \, \mathrm{m \, s^{-1}}$, and therefore this upward pressure gradient must be close to balance with the gravitational force acting towards the centre of the Earth. This balance between the vertical pressure gradient and the gravitational force is known as hydrostatic equilibrium. Ideally, the state requires no motion and therefore the atmosphere is never quite in hydrostatic balance. However, the approximation is very good for large scales of motion and only breaks down in regions of large vertical acceleration associated with thunderstorm updrafts and mountain waves.

Consider a slab of atmosphere of thickness δz and cross-section area δA. The pressure force across the slab is $[p - (\partial p / \partial z) \delta z] \delta A - p \, \delta A$:

$$\mathrm{PF} = -\frac{\partial p}{\partial z} \delta z \delta A$$

The force due to gravity on the element of atmosphere is $\rho \, \delta A \, \delta z g$, hence the balance is $(\partial p / \partial z) \, \delta z \, \delta A = -\rho \, \delta A \, \delta z g$ or

$$\frac{\partial p}{\partial z} = -\rho g \qquad (2.3)$$

The hydrostatic equation 2.3 is a very important relationship in both meteorology and oceanography because it relates the pressure to the depth of fluid. In the atmosphere, however, the density also varies with height because of the compressibility of gases, and therefore equation 2.3 is not sufficient. Since the atmosphere is a very good approximation to an ideal gas, the ideal gas equation can be used to eliminate density in favour of temperature and pressure. On substitution of $\rho = p/RT$, we have

$$\frac{\partial p}{\partial z} = -\frac{pg}{RT}$$

If T is replaced by the vertical mean temperature of the atmosphere, $\bar{T}$, the equation can be integrated from the surface ($z = 0$) to yield the pressure at a height z',

$$\int_{p0}^{p'} \frac{\partial p}{p} = -\frac{g}{R\bar{T}} \int_0^{z'} \partial z$$

or

$$p' = p_0 \exp(-z'/H) \qquad (2.4)$$

where p_0 is the surface pressure and $H = R\bar{T}/g = R^*\bar{T}/g\bar{m}$. This equation is sometimes known as the barometer equation. The quantity H is known as the scale height and is a measure of the depth of the atmosphere. When $z' = H$, then $p' = p_0 e^{-1}$ or $p' = 0.37 p_0$. Therefore, H is the height at which the pressure is 37% of the surface atmospheric pressure. Furthermore, the pressure decreases by $1/e$ for each increase in height equal to the scale height.

Table 2.3 gives typical scale heights for the Earth's atmosphere and other planetary atmospheres, assuming the mean temperature is equal to the radiative equilibrium temperature. It is noted that the scale height is related to mean molecular weight of the atmosphere and the acceleration due to gravity. On Mars, for instance, the gravity is approximately one-third of the Earth's gravity, but the atmosphere is composed of carbon dioxide with a molecular weight of 44, and hence there is only a small increase in scale height. On Venus, the large vertical variation in temperature makes the scale height a highly variable parameter. The scale height is largest in the lowest layers of the Venusian atmosphere (Table 2.3), where the temperature approaches 760 K (Table 1.1).

As the reader will appreciate, the scale height is not a constant for a given atmospheric composition because of variations in the temperature of the atmosphere. To obtain accurate results from the equation, it is necessary to measure the temperature profile of the atmosphere as a function of height, and then to integrate the equation numerically. A simpler alternative to this procedure is to use a standard atmospheric temperature profile to obtain the pressure–height relationship. This was used to calibrate pressure altimeters for aircraft, but these have been superseded by the more accurate radar altimeters.

In the ocean, the relationship between pressure and depth is very close to linear because of the small variation of density with depth. In the surface layers of the ocean, to a depth of 1 km, it is normal practice for the compressibility of sea water to be ignored. However, because of the exceptionally high pressures ($10^4 - 10^5$ kPa) existing in the deep water, compressibility does become an important effect when considering deep-water masses. An illustration of the importance of compressibility is that sea level would be 30 m higher than it is today if the ocean was assumed to be incompressible.

The pressure at a depth h in the ocean is obtained by integration of equation 2.3 from the surface to the depth h. Strictly speaking, this is not as easy as you might think, because the surface of the ocean varies in height as the result of tides and large-scale water movements. The variations in sea level may be as large as 1 m. Secondly, the surface atmospheric pressure can vary by as much as 10% of the mean surface pressure, which will produce horizontal pressure variations in the ocean. However, for the calculation of total pressure in the ocean, these effects are relatively small and may often be neglected, although in Chapter 7 it will be shown that these small changes in pressure are important in determining horizontal currents. As a general rule of thumb, 10 m of sea water is equivalent to a pressure of 10^5 Pa (or 1 atm).

2.5   Adiabatic changes and potential temperature

Consider a balloon which is thermally insulated, so that no heat is lost through the surface of the balloon, and the balloon is filled with hydrogen and released into the atmosphere. At the moment of release, it is assumed that the hydrogen has the same temperature as the surrounding air. As the balloon ascends, it will expand because of the reduction of atmospheric pressure with height and, at the same time, the hydrogen in the balloon will become cooler. This reduction in temperature is solely due to reduction in the internal energy of the gas, as a result of energy used to expand the surface of the balloon. Since no heat is gained or lost to the hydrogen gas, by thermal conduction across the surface or by thermal radiation from the atmosphere or Sun, the balloon is known as an adiabatic system and the reduction in temperature is known as adiabatic cooling. Of course, if there is no transfer of energy to its surroundings and the balloon is brought back down to the surface at the same initial pressure, then the temperature of the hydrogen gas will be the same as the initial temperature. Therefore, a property of an adiabatic system is that the system is reversible and, therefore, no matter which path the balloon takes through the atmosphere when it returns to its initial pressure level, it will have the same initial temperature.

These properties of the adiabatic system can be understood by the application of the first law of thermodynamics:

$$dQ = C_v \, dT + p \, d\alpha \qquad (2.5)$$

$$\underset{\substack{\text{heat added} \\ \text{to the system}}}{} \quad \underset{\substack{\text{internal} \\ \text{energy}}}{} \quad \underset{\substack{\text{work done} \\ \text{by expansion}}}{}$$

The first term represents the heat added across the boundaries of the system, which could be associated with heat conduction across the balloon's surface and radiational heating. The second term is the internal energy of the gas, which is associated with the average kinetic energy of the molecules and proportional to the temperature of the gas. C_v is the specific heat of the gas at constant volume. The third term is the work done on the surroundings and, in our example, is associated with expansion of the volume of gas in the balloon. Hence, heat added to the system may either cause a change in volume of the gas or it may result in a change in temperature of the gas, or it may be a combination of both effects. As mentioned earlier, for an adiabatic system there is no heat added across the boundaries of the system, and therefore $dQ = 0$. Hence, it can be seen from equation 2.5 that an expansion of the balloon ($d\alpha > 0$) will cause a reduction in the internal energy of the gas and a decrease in temperature.

Before this relationship can be usefully used, it is necessary to substitute for the change in specific volume, $d\alpha$. By differentiating the ideal gas equation by parts, it can be shown that $\alpha\,dp + p\,d\alpha = R\,dT$. By substituting for $p\,d\alpha$ in equation 2.5 and noting $C_p = C_v + R$, where C_p is specific heat at constant pressure, then $0 = C_p\,dT - \alpha\,dp$; hence:

$$\frac{dT}{dp} = \frac{\alpha}{C_p} = \frac{1}{\rho C_p}$$

but, since ρ depends on pressure, this is not a particularly useful relationship. However, by rewriting the equation as

$$\frac{dT}{dz}\frac{dz}{dp} = \frac{1}{\rho C_p}$$

and substituting for the hydrostatic equation (equation 2.3), we have:

$$\frac{dT}{dz} = -\frac{g}{C_p} \tag{2.6}$$

The variation of temperature with height in an adiabatic system is therefore determined solely from the acceleration due to gravity and the specific heat of the gas at constant pressure. For the lower atmosphere (<10 km), the change in g with height is small, and therefore the adiabatic temperature gradient is, to a good approximation, a constant. Table 2.5 shows calculated adiabatic temperature gradients for the Earth and other planets. It is noted that both Earth and Venus have similar adiabatic temperature gradients, despite the different composition of their atmosphere and masses. It is necessary to emphasize that the adiabatic temperature gradient is only

Table 2.5 Adiabatic temperature gradients for planetary atmospheres

	Earth	Mars	Venus	Jupiter
Adiabatic temperature gradient (K km^{-1})	9.8	4.5	10.7	20.2

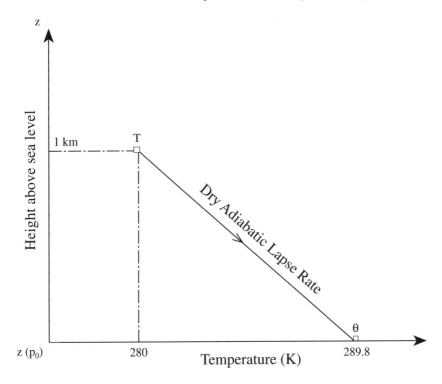

Figure 2.5 Relationship between temperature and potential temperature in the atmosphere as a function of height above mean sea level. p_0 is mean sea-level pressure

obtained when parcels of air do not gain or lose heat to their environment. In practice, mixing of air parcels, energy interchange by radiation and the condensation of water vapour produce a change in total energy of the air parcel. The latter two effects are known as non-adiabatic processes or diabatic processes. Because of the diabatic processes, environmental temperature profiles are rarely adiabatic, though we shall see that the concept is still a useful one in understanding the vertical stability of the atmosphere. The average observed temperature gradient in the lowest 10 km of the atmosphere is $\sim$6.5 K km^{-1}, compared with an adiabatic temperature gradient of 9.8 K km^{-1}.

The adiabatic temperature gradient for the ocean can also be derived from equation 2.5, but the final expression is not as simple because of

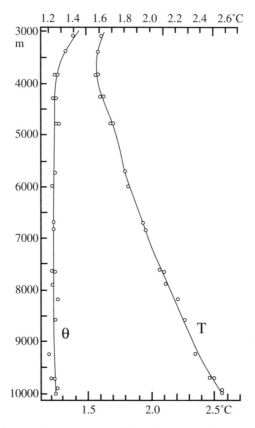

Figure 2.6 Observed temperature (T) and potential temperature (θ) in the Philippines Deep (Mindanao Trench). Observations from the Danish vessel *Galathea* in July–August 1951

the more complicated form of the equation of state. In this case, equation 2.5 becomes

$$dQ = C_p \, dT - T \left(\frac{\partial \alpha}{\partial T} \right)_p dp$$

Assuming $dQ = 0$ and substituting for the hydrostatic equation, it can be shown that $dT/dz = -T\beta g/C_p$, where $\beta = (1/\alpha)(\partial \alpha/\partial T)_p$. β is the coefficient of heat expansion of sea water at constant pressure.

Unfortunately, both β and C_p for sea water are dependent on pressure, temperature and salinity. At atmospheric pressure, standard ocean salinity of 35 psu and a temperature of 0°C, the adiabatic temperature gradient is $-0.035\,\mathrm{K\,km}^{-1}$, whilst at a pressure of 500 atm (equivalent to a depth of 5000 m), the adiabatic temperature gradient is $-0.1\,\mathrm{K\,km}^{-1}$. Thus, the adiabatic temperature gradient in the ocean is $\leq 1\%$ that of the atmospheric adiabatic temperature gradient. However, because of the small variation of temperature in the deep-water masses, the adiabatic temperature gradient has to be taken into account when considering water masses at different depths.

We have seen that the adiabatic effect of pressure on temperature is significant in both atmosphere and ocean, and therefore any vertical motions will produce an adiabatic temperature change in a parcel of fluid. Also, we have indicated that radiation, mixing and water vapour condensation (in the atmosphere) will also produce significant changes in temperature of a parcel of air or sea water. It is useful, therefore, to separate the adiabatic process from diabatic processes in order to distinguish the two effects. For instance, a meteorologist may wish to decide if the air mass at the surface is similar to the air mass observed in the upper atmosphere. Similarly, an oceanographer may wish to determine the similarity of water masses at two different depths in the ocean. For these processes, it is necessary to remove the adiabatic effect by bringing the air mass or water mass (adiabatically) to a reference pressure. This reference pressure is $10^5\,\mathrm{Pa}$ for the atmosphere and surface atmospheric pressure for the ocean. Because of the small adiabatic effect in the ocean, the effect of variations in atmospheric pressure are negligible. The temperature of the fluid parcel at the reference pressure is known as the potential temperature, θ. Figure 2.5 shows the temperature change of an air parcel moved adiabatically a vertical distance of 1 km from the level of the reference pressure. The temperature at the initial position is 280 K, but the potential temperature is 289.8 K. Generally, since the reference pressure is close to sea-level pressure, the atmosphere potential temperatures are usually larger than the observed temperatures. In the ocean, the reference pressure is the minimum pressure of the ocean (i.e. surface), and therefore ocean potential temperatures are always lower than *in situ* temperatures. Figure 2.6 shows the importance of reducing *in situ* temperatures to potential temperatures in the deep ocean. The measurements show temperature increasing steadily towards the bottom in the Mindanao Trench, which suggests a strong vertical instability of the lower layer (assuming salinity is not stabilizing the

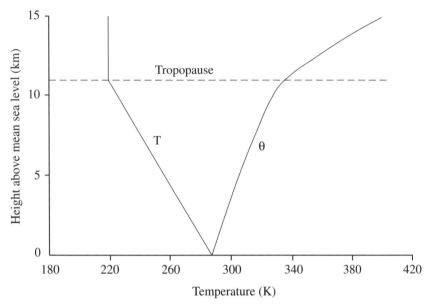

Figure 2.7   Mean vertical profile temperature (*T*) and potential temperature (*θ*) in the lowest 15 km of the atmosphere

temperature gradient). However, the potential temperature is in fact nearly constant with depth, rather than increasing with depth. The apparent warm bottom layer is, therefore, solely an artefact of the adiabatic effect.

An example of the reduction of an observed atmospheric temperature profile to a profile of potential temperature is given in Figure 2.7. It is again noted that the removal of the adiabatic influence substantially changes the profile, from one which decreases with height to a profile where the potential temperature increases with height. Usually, it is only close to the surface under strong heating conditions that a negative gradient of potential temperature exists. In the free atmosphere generally, potential temperature increases with height. The reason for this will be discussed in the following section.

2.6   Vertical stability of the ocean and atmosphere

We will first consider the stability of an incompressible fluid in the two situations shown in Figure 2.8b and d. Initially, the parcel of fluid at

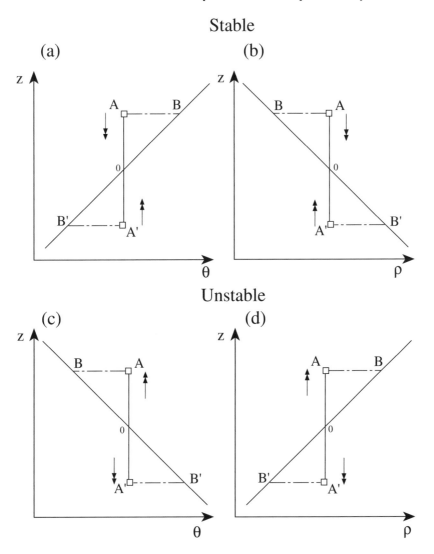

Figure 2.8 Stability of a parcel perturbed from equilibrium for atmosphere (a, c) and ocean (b, d)

point 0 is in hydrostatic equilibrium with its environment (i.e. the upward pressure gradient on the parcel is balanced by the downward force due to gravity). If the parcel of fluid in case (b) is moved *upwards* by an external force to a position *A*, and provided there is no mixing of the parcel with its environment, then the parcel will have a density greater than its environment density at that level. The parcel is now heavier than its environment and will therefore accelerate back towards its original level. In this case we say that the fluid system is *stable*. However, in case (d), the parcel at *A* will have a density less than the environment density at that level. In this case, the fluid is lighter than its surroundings and the parcel will be subjected to an upward buoyancy force, which will accelerate the parcel away from its original position. Furthermore, as the parcel moves further from its initial position, it will be subjected to a larger buoyancy force, and therefore will be subjected to larger accelerations. In this case the fluid system is *unstable*. It should be appreciated that downward displacements of the parcels from the initial position will behave in a similar manner. In the stable case the fluid parcel will move back to its original position, whilst in the unstable case it will accelerate away from its original position. In the situation where the density is uniform throughout the fluid, there will be no buoyancy forces and the system is said to be neutral. In this case, parcels will continue to move in their initial direction until frictional forces slow them down.

The above criterion for stability can be applied to both ocean and atmosphere, provided the adiabatic effects are taken into account. We have shown that a conservative property of an adiabatic fluid is the potential temperature and, in the atmosphere, the criterion for stability is simply based on the vertical gradient of potential temperature. If the potential temperature *increases* with height, the atmosphere is *stable* (Figure 2.8a); whilst if it *decreases* with height, it is *unstable* (Figure 2.8c). It is left to the reader to consider the buoyancy forces on parcels displaced from equilibrium for these three cases, i.e.

$$\text{stable} \quad \frac{\partial \theta}{\partial z} > 0$$

$$\text{neutral} \quad \frac{\partial \theta}{\partial z} = 0 \qquad (2.7)$$

$$\text{unstable} \quad \frac{\partial \theta}{\partial z} < 0$$

Figure 2.9 shows the diurnal variation in potential temperature in the lowest 1 km of the atmosphere over land. During the day-time, the potential temperature is uniform up to the height of the convection and the potential temperature gradient is stable above this level. Only very

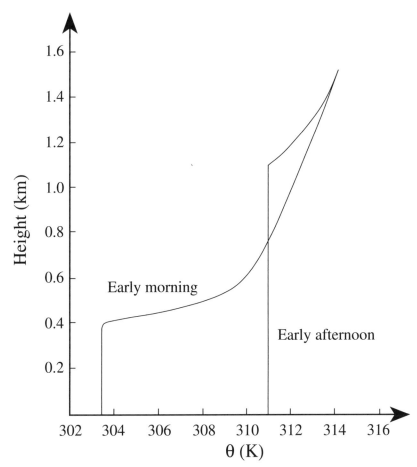

Figure 2.9 Diurnal variation in potential temperature in the atmospheric boundary layer over land in summer

close to the surface can a negative gradient of potential temperature exist. The reason for this is that the radiational heating of the surface can cause a higher surface temperature than air temperature and sustain a negative gradient close to the surface. However, away from the surface, large convection currents driven by the unstable temperature gradient can mix the boundary layer very effectively, and thus a uniform potential temperature (i.e. $\partial\theta/\partial z = 0$) is produced. As midday approaches, the surface heating reaches a maximum, and the convection currents become deeper and mix stable air into this boundary layer. Therefore, the region of uniform potential temperature (or the mixed layer) increases in depth. During the night, the ground cools by loss of planetary radiation more effectively than the air above it, and therefore the potential temperature gradient becomes positive and the air becomes stable. In those circumstances the air is quiescent, and smoke and other pollutants will tend to be trapped close to the ground. Only by artificially adding buoyancy to the pollutants can they be dispersed effectively.

In the ocean, the situation is complicated by the effect of salinity, hence the vertical gradient of potential temperature does not always indicate the stability of the ocean. Generally, we have to use a quantity called the potential density (σ_θ), which is the density a parcel of sea water would have if it was moved adiabatically to the sea surface. For deep-water masses, a reference level of 4 km is used.

The stability of the ocean is then determined by the vertical gradient of potential density, i.e.

$$\text{stable} \quad \frac{\partial \sigma_\theta}{\partial z} < 0$$

$$\text{neutral} \quad \frac{\partial \sigma_\theta}{\partial z} = 0 \tag{2.8}$$

$$\text{unstable} \quad \frac{\partial \sigma_\theta}{\partial z} > 0$$

Unstable density gradients are very rarely observed in the ocean. The reason for this is that the convective instability process is very efficient in mixing the unstable density gradient to produce a neutral profile. The presence, therefore, of a well-mixed ocean layer is a good indication of a convective instability process. The principal cause of the convective instability in the ocean is strong cooling of the surface during winter, which causes the formation of a relatively dense surface layer which will be unstable with respect to the deeper water and cause overturning. The surface cooling has a dual effect in destabilizing the water. It will not only remove heat from the water column, but increased evaporation will tend to increase the surface salinity and thus increase the density. The depth to which convection occurs will depend on the vertical density structure and the intensity of the surface cooling. In the North Pacific, the depth of winter convection is limited to the upper 150 m because of the presence of a very stable halocline below this depth. However, in the Labrador Sea, deep convection has been observed to occur to a depth of 1500 m and, in the western Mediterranean in the Gulf of Lyons, deep convection occurs to the bottom.

Buoyancy frequency

Let us reconsider the way in which a parcel of fluid returns to its equilibrium position (Figure 2.8b and d) in a stable system. If there is no mixing of the parcel with its environment and no frictional drag on the parcel, then the parcel will accelerate towards its equilibrium position and overshoot. It will then find itself in an environment of higher density and it will accelerate in the opposite direction back towards its equilibrium position. The parcel will therefore oscillate about its equilibrium position with a frequency related to the magnitude of the density gradient. If the density gradient is large, then the buoyancy accelerations will be correspondingly increased and the frequency will also increase. This frequency is called the Brunt–Väisälä frequency or the buoyancy frequency (N).

Consider the buoyancy force on a parcel displaced a distance δz from its equilibrium density ρ_e. The buoyancy force is $-(\rho_e - \rho)g$. If $\rho = \rho_e + (\partial\rho/\partial z)\,\delta z$, then $(\rho_e - \rho)g = -g(\partial\rho/\partial z)\,\delta z$. The acceleration as a result of the buoyancy force is $dw/dt = g/\rho(\partial\rho/\partial z)\,\delta z$. Since $w = d z/dt$, then

$$\frac{d^2}{dt^2}\,\delta z = -\left(-g\frac{1}{\rho}\frac{\partial\rho}{\partial z}\right)\delta z$$

This equation is analogous to the simple harmonic equation $d^2x/dt^2 = -N^2x$, where N is the frequency. Hence, $N = \sqrt{[-(g/\rho)(\partial\rho/\partial z)]}$. Applying this expression to the atmosphere, the density is replaced by potential temperature, i.e.

$$N = \sqrt{\left(\frac{g}{\theta}\frac{\partial\theta}{\partial z}\right)} \quad \text{where} \quad \frac{\partial\theta}{\partial z} \geq 0$$

for the ocean

$$N = \sqrt{\left(\frac{-g}{\sigma_\theta} \frac{\partial \sigma_\theta}{\partial z} \right)} \quad \text{where } \frac{\partial \sigma_\theta}{\partial z} \geq 0 \qquad (2.9)$$

where σ_θ is the potential density.

The buoyancy frequency N provides a useful way of describing the stability of both the atmosphere and the ocean, and it is a very important parameter in the advanced study of dynamical oceanography and meteorology. One of its most important uses is the determination of the frequency of internal waves, which are continually produced by turbulent motions in both systems. In the ocean thermocline, N has a period of 10 min, whilst in the deep ocean, where the stability is weaker, it has a typical period of 50 hours. Internal waves in the thermocline, produced by surface storms, can have amplitudes of 10–50 m. In the atmosphere, the period of the waves is shorter than that of waves in the ocean, ranging from 30 s to 10 min. The more spectacular internal waves are produced by forced ascent over mountains, which in certain circumstances can produce wave trains (lee waves) extending for up to 100 km downstream of the mountain. The waves are most apparent when the wave crests are above the condensation level.

3

Radiation, Temperature and Stability

3.1 Vertical variation of atmospheric constituents

The vertical variations of constituents in the atmosphere are important in understanding the way in which solar and planetary radiations are absorbed in the atmosphere, and the resulting effect on the temperature profile of the atmosphere. Here, the physical processes which control the distribution of atoms and molecules in the atmosphere will be considered. First, consider the major constituents of the atmosphere: oxygen, nitrogen, argon, water vapour and carbon dioxide. If all these constituents were initially mixed together and left for a long time in an undisturbed environment, the distribution of each constituent would vary with height according to its molecular or atomic weight. The lighter molecules would have a smaller variation with height than the heavier molecules, and thus the lighter molecules would be more abundant in the upper atmosphere than in the lower atmosphere compared with the heavier molecules. The equilibrium distribution is given by the hydrostatic equation for each constituent gas rather than for the mixture of gases. Therefore, it can be seen that the rate of change of partial pressure (P_n) of each gas is directly proportional to the molecular weight (m_n) of the gas, i.e.

$$\frac{\partial P_n}{\partial z} = -\frac{m_n P_n g}{R^* T}$$

However, the observed distribution of the gases in the atmosphere does not show this relationship. All of the gases, even carbon dioxide which is the heaviest, are well mixed to a height of 100 km and it is only above this level that evidence of gravitational settling is apparent. To understand this distribution, it is necessary to consider the rate at which diffusion occurs. The rate of gravitational settling of gases is determined by the molecular diffusion rate for each gas. The diffusion rate is defined by the product of the velocity of the molecules and their mean free path between collisions. The velocity of the atom or molecule increases with the temperature of the gas and decreases with the molecular weight. Hence, the diffusion rate is higher for light gases at low pressures and high temperatures than for heavier gases at low temperatures and high pressures. There is a rapid decrease of pressure with height in the atmosphere and, because of this, the most significant factor in the determination of diffusion rates is the mean free path. The diffusion rate is many orders of magnitude higher in the upper atmosphere, where the mean free path is very large, than in the lower atmosphere. Typically, at a height of 120 km, it would take approximately one day for gravitational separation to occur but, for the atmosphere taken as a whole, the typical time scale is approximately 30 000 years. The atmosphere is always in a state of turbulent motion, and therefore the molecules are being continually mixed by a variety of atmospheric motions. The rates of mixing are much larger than the corresponding rates of gravitational settling in the lower atmosphere, and therefore the profiles of these constituents are well mixed. Only when the gravitational settling time becomes of the same order as the mixing time scale is diffusive separation observed. For molecular oxygen and nitrogen, this occurs typically at a height of 100 km. Above these levels the gases are distributed according to their molecular weights, and therefore hydrogen and helium occur in increasing abundancies at high levels, whilst the abundancies of the heavier gases, such as oxygen and nitrogen, decrease with height.

It is only for the noble gases that the story finishes here. For all the other constituents, their interaction with radiation has to be considered when discussing their vertical distribution in the atmosphere. These interactive processes include:

(i)   The dissociation of a molecule by radiation.
(ii) The combination of atoms by collisions.

The dissociation energy of a molecule is the energy required to break the molecular bond. The energy of one photon of radiation is given by

$h2\pi/\lambda$, where h is Planck's constant and λ is the wavelength of the photon, and it is therefore accepted practice to define the dissociation energy in terms of the wavelength of the radiation. For an oxygen molecule the dissociation energy corresponds to a wavelength of $0.24\,\mu\mathrm{m}$, and therefore all radiation having a smaller wavelength, and hence a higher energy, than this will dissociate the oxygen molecules into two atoms of oxygen, i.e.

$$O_2 + h\upsilon \rightarrow O + O, \quad \lambda \leq 0.24\,\mu\mathrm{m}$$

Nitrogen has a higher binding energy than oxygen and only photons having wavelengths less than $0.13\,\mu\mathrm{m}$ are able to dissociate the nitrogen molecule.

Atomic oxygen becomes an increasingly abundant constituent above a height of 20 km, but it is only above 120 km that molecular oxygen is replaced by atomic oxygen as a dominant oxygen constituent. The reason for this distribution is that the rate of molecular dissociation is proportional to the product of the number of oxygen molecules and the number of photons having more than the required energy for dissociation. At very high elevations there are a large number of photons with the energy necessary to dissociate oxygen molecules but, due to gravitational settling and dissociation, there are few oxygen molecules, and therefore the rate of production of atomic oxygen is low. However, in the lower atmosphere, though there is an abundance of molecular oxygen, there are only a few photons possessing the required energy to dissociate the molecules, because many of these photons have been absorbed in the upper levels of the atmosphere. Therefore, it is only in the middle levels, around 100 km, where there are sufficient molecules and photons to produce a large dissociation rate, and hence to produce a high number of oxygen atoms. However, the predicted height of maximum atomic oxygen production does not agree with the observed distribution. This is due to two additional processes in the lower atmosphere which have so far been neglected and which act to reduce the number of oxygen atoms. The first process is the recombination of atomic oxygen into molecular oxygen by collision. This process is very effective at higher pressures and it is therefore a major sink of atomic oxygen at lower altitudes, between 70 and 100 km. The second process is the production of ozone, O_3, from a three-way collision between an atom of oxygen, a molecule of oxygen and another unspecified molecule, M:

$$O + O_2 + M \rightarrow O_3 + M$$

The unspecified molecule is necessary because it absorbs the energy released when a molecule of ozone is formed. The production of ozone depends on the relative numbers of oxygen molecules and atoms, and the number of three-way collisions, which depends in turn on the total pressure of the gas. For a photochemical equilibrium, the rate of production of atomic oxygen must be equal to its rate of destruction and, given the photoabsorption properties of the constituents, it is possible to obtain the vertical equilibrium distribution. This profile is known as the Chapman profile.

Figure 3.1 shows the relative distributions of atomic oxygen, molecular oxygen and ozone in the atmosphere. It is noted that, because the ozone distribution is dependent on both the collision frequency and the amount of atomic oxygen, it tends to have its highest production rate between 30 and 100 km. Below 30 km, the infrequency of occurrence of atomic oxygen reduces the rate of ozone formation to insignificant levels. Above 100 km, it is the low collision frequency which brings about the reduction of the rate of ozone formation. However, there are three major processes which result in the destruction of ozone:

(i)   The photodissociation of ozone which occurs for photon wavelengths less than 11 μm, which therefore operates throughout the solar and planetary spectra

$$O_3 + h\upsilon \rightarrow O_2 + O, \quad \lambda \leq 11\mu m$$

The strongest dissociation occurs for wavelengths between 0.2 and 0.3 μm, and therefore ozone is a very effective absorber of ultraviolet radiation.

(ii) The recombination of ozone with atomic oxygen

$$O_3 + O \rightarrow 2O_2$$

(iii) The removal of ozone at the Earth's surface by oxidation.

It can be seen that, at high levels of the stratosphere, ozone will be readily destroyed by processes (i) and (ii), whilst process (iii) will destroy ozone in the troposphere. In the troposphere, mixing rates are larger

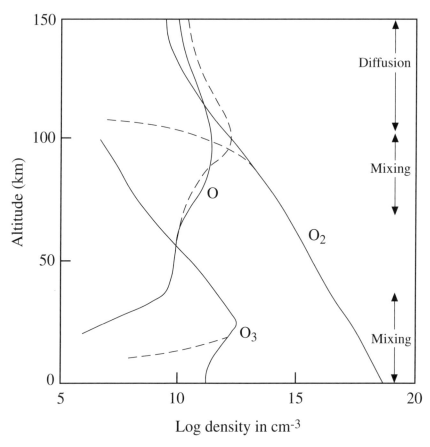

Figure 3.1   Vertical profiles of the densities of atomic oxygen (O), molecular oxygen (O_2) and ozone (O_3) in the Earth's atmosphere. Broken lines show the theoretical profile calculated on the assumption of photochemical equilibrium

than in the stratosphere where most ozone is formed, and therefore ozone molecules entering the troposphere are brought to the surface in a matter of a few days and oxidation then takes place. Hence, ozone tends to have a concentration maximum in the lower stratosphere where destructive processes are at a minimum. Here, individual ozone molecules may have lifetimes of the order of a few months. The detailed photochemistry of ozone is further complicated by other trace gases, such as nitrous oxide (which is related to biological production processes at the surface and vehicle exhausts) and chlorofluoromethanes (present as aerosol propellants in many sprays), and other man-made gases. All of these gases, in certain circumstances, can act to reduce the level of ozone in the atmosphere, and hence reduce the ultraviolet shielding at present enjoyed by the Earth. In the last decade, the link between increasing levels of chlorine in the stratosphere and the decrease in ozone there has become well established.

The interrelationship between atomic oxygen, molecular oxygen, ozone and radiation is an example of the complexity of photochemical processes which produce the observed distributions of the constituent gases of the atmosphere. For the other principal gases, water vapour and carbon dioxide, dissociation is only significant at altitudes greater than 100 km. Carbon dioxide is dissociated by ultraviolet radiation of wavelengths less than 0.17 μm and water molecules commence dissociation at 0.24 μm. The vertical distribution of water vapour in the atmosphere below 100 km is determined principally by the condensation process. The rapid decrease of pressure and temperature with height in the lowest 20 km of the atmosphere results in a reduction of the saturated mixing ratio (defined as the ratio of the mass of water vapour in a given volume of air, saturated with respect to a plane water surface, to the mass of dry air in the same volume) from 10^{-2} at the surface to 10^{-5} at 20 km. Therefore, for every tonne of dry air in the stratosphere, there is only 10 g of water vapour. The temperature minimum at the tropopause, 10 km above the surface, acts as a cold trap for the movement of water vapour between the troposphere and the stratosphere. However, the water vapour in the upper atmosphere is still an important constituent when considering the absorption of solar and planetary radiation. At 80 km, where the temperature reaches a second minimum at the mesopause, occasionally there is sufficient water vapour to allow the formation of thin, noctilucent ice clouds. These clouds cannot be observed during the day, but are often seen just after sunset in high latitudes when the reflected sunlight from the clouds may be observed at the surface.

The reasons for the temperature minima at the tropopause and mesopause will be discussed in Section 3.4.

3.2   The attenuation of solar radiation

The observation that the maximum in the solar-radiation spectrum occurs at 0.5 μm, whilst planetary radiation peaks between 10 and

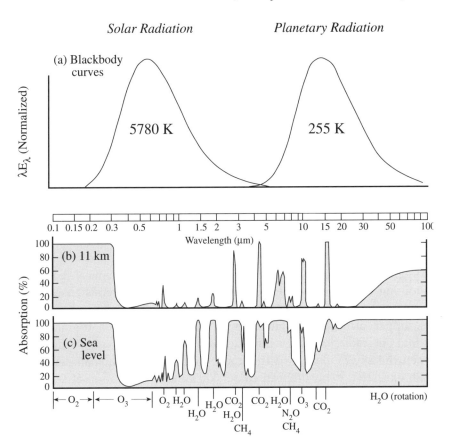

Solar Radiation   *Planetary Radiation*

Figure 3.2 Attenuation of solar radiation. (a) Normalized blackbody curves for 5780 K and 255 K, plotted so that irradiance is proportional to the areas under the curves. (c) Atmospheric absorption in clear air for solar radiation with zenith angle of 50° and for diffuse planetary radiation. (b) Same as for (c) but for the portion of the atmosphere lying above the 11 km level, near the middle-latitude tropopause

15 μm (in the middle infra-red), allows separation of the radiation processes due to solar-radiation absorption from those involved in planetary-radiation absorption. Only 0.4% of the total solar radiation has a wavelength longer than 5 μm, whilst a similar small percentage of planetary radiation from a surface at a temperature of 288 K has a wavelength less than 5 μm. The propagation of solar radiation through the atmosphere is simpler to understand than that of planetary radiation, because no re-emission of radiation occurs as the solar

radiation travels through the atmosphere, and therefore it is only necessary to consider the attenuation factors. The situation with respect to the planetary-radiation spectrum is complicated by the re-emission of radiation at similar wavelengths to the absorbed radiation. This will be considered in Section 3.3.

The attenuation of solar radiation in a dust-free and pollution-free atmosphere with no clouds (Figure 3.2) would be the result of two processes:

(i) Gaseous absorption.
(ii) Molecular scattering.

The primary gaseous absorption occurs in the ultraviolet region of the spectrum, at wavelengths less than $0.3 \, \mu$m. As previously discussed, this absorption mainly results in the photodissociation of molecular oxygen and nitrogen above 100 km, and the photochemical reactions involving ozone in the stratosphere. This primary gaseous absorption results in a nearly complete removal of lethal ultraviolet radiation by the upper atmosphere. However, a small fraction of the incident ultraviolet radiation does manage to reach the surface, particularly when the sky is clear, and this can be damaging to human skin and, with excessive exposure, can cause skin cancers. Radiation at $0.26 \, \mu$m damages the DNA molecule which is present in all living organisms. In quantitative terms, the amount of ultraviolet radiation absorbed is less than 2% of the total solar flux for wavelengths less than $0.32 \, \mu$m. Ozone has absorption bands in the visible region of the spectrum, but these bands are generally weaker than the ultraviolet absorption bands. The only other significant absorber of solar radiation is water vapour, which has a number of absorption bands in the near infra-red part of the solar-radiation spectrum. The absorption bands of water vapour have a very significant quantitative effect on the amount of solar radiation actually reaching the surface, because of the large amount of solar radiation in this region of the spectrum, and this effect is illustrated in Figure 3.2. The variability of the amount of water vapour present in the atmosphere means that the effect of its infra-red absorption will vary over the globe.

Molecular scattering is proportional to λ^{-4}, where λ is the wavelength of the incident photon. Thus, the shorter-wavelength light is scattered rather more than the longer-wavelength light. For example, blue-light scattering is four times greater than red-light scattering, and hence the

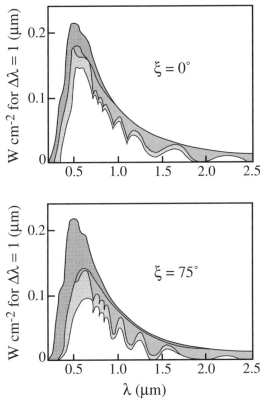

Figure 3.3 Molecular scattering, absorption and scattering by aerosols for a solar zenith angle of 0° (upper) and 75° (lower). The upper curve shows the spectral distribution of the solar constant, while the lower curve represents the spectral distribution of the direct solar beam at the sea surface, for clear dry weather (equivalent to water thickness equal to 1 cm). The areas of oblique lines, cross-hatching and dots represent, respectively, the effects of molecular scattering, absorption and scattering by aerosols

direct solar beam loses a significant fraction of its blue light. On a clear day, with a dust-free atmosphere, the shortest wavelength radiation is seen at the antisolar point, i.e. the point at 180° to the direction of the Sun and at the same elevation as the Sun. This part of the sky often has a purple hue and ultraviolet radiation is at a maximum from this direction.

The presence of small suspended particles, of diameter generally less than a few micrometres, in the atmosphere has a significant influence on the solar radiation received at the surface. Such particles, called the aerosol, can both scatter and absorb radiation. The scattering of solar radiation tends to increase the path length of a photon through the atmosphere, and therefore it leads, indirectly, to an increase in the probability of absorption. The scattering by the aerosol varies as λ^{-1}, i.e. it is not so strongly wavelength dependent as molecular scattering, and therefore it is significant throughout the solar spectrum. From Figure 3.3, it may be seen that the attenuation by the aerosol increases as the elevation of the Sun decreases. This is the result of the increased path length through the atmosphere and, therefore, the increased chance that a photon will be absorbed. Hence, in polar latitudes in summer, an aerosol layer will tend to produce enhanced absorption.

There are many types of aerosol in the atmosphere, but a broad classification into major types is possible. Considering the Earth as a whole, the natural aerosol dominates over the man-made aerosol, although the latter is dominant in highly industrialized regions such as Europe and the eastern USA. The natural aerosol is produced by wind turbulence at the Earth's surface. Over land the particles are mainly silica and clay, and over the ocean they are mainly salt. In addition, volcanic activity can inject large quantities of sulphur particles into the troposphere and the lower stratosphere. In the El Chichon eruption in 1982, it is estimated that 10^7 tonnes of material was injected into the stratosphere and, in Hawaii, the direct solar radiation was reduced by approximately 7% following the explosion. The aerosol produced by El Chichon was traced across the Pacific Ocean for many months. Other sources of the natural aerosol include forest fires which can inject large quantities of carbon particles into the troposphere. These carbon particles are particularly effective in absorbing solar radiation. An important type of aerosol is formed from hygroscopic substances such as sodium and chloride. These aerosols act as very effective cloud condensation nuclei, and hence maritime salt haze becomes particularly effective in reducing solar radiation when the humidity reaches approximately 95%.

Human activities increase the input of sulphur into the troposphere in the form of sulphur dioxide and ammonium sulphates, and the effect of this input is important both in the formation of acidic rain and its influence on solar radiation. The carbon input from coal- or oil-based industries can also produce very significant reductions in the solar radiation. A dramatic example was the ignition of oil wells in Kuwait

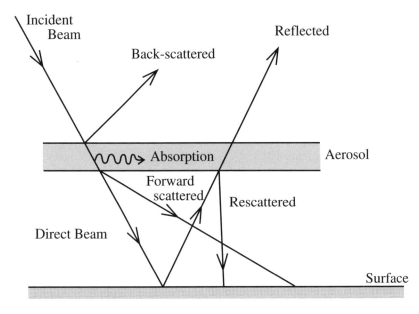

Figure 3.4 Schematic diagram of scattering and absorption by an atmospheric aerosol layer

following the Gulf War (1991). This produced a 5 km-deep plume of smoke which extended for many hundreds of kilometres downwind. In the centre of the plume, the solar radiation was reduced to zero and the surface temperature dropped by 10°C. Again, as with the salt haze, as the humidity increases condensation occurs on hygroscopic pollutants, such as ammonium sulphate, and this increases their influence on the solar radiation.

Figure 3.4 shows how the aerosol may affect the path of a photon through the atmosphere. The back-scattering of incident photons from the aerosol will tend to enhance the albedo of the atmosphere. The photons which are not back-scattered into space, in the absence of gas absorption or clouds, will be either absorbed by the aerosol or reach the Earth's surface. Some of the photons will be reflected back into the atmosphere by the surface. A fraction of the reflected photons will be back-scattered to the surface by the aerosol and the remaining photons will be lost to space. The relative magnitudes of each fraction will depend on the absorption to back-scattering ratio of the aerosol and the surface albedo. Although these properties depend on the type of

(a)

(b)

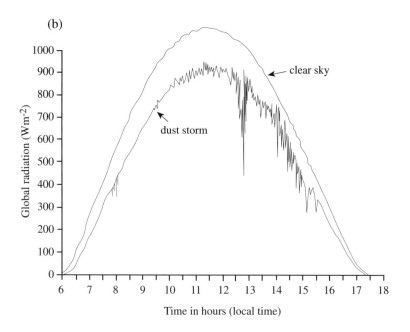

aerosol and the type of surface, some broad trends have been identified. First, if the surface albedo is less than 30%, then the aerosol will tend to increase the planetary albedo. Surface albedos of such magnitudes are typical of all ocean surfaces and most land surfaces, except for sand deserts. If the surface has a higher albedo, light trapping between the surface and the aerosol will tend to decrease the planetary albedo. Secondly, the absorption to back-scattering ratio varies between 1 and 20% for maritime 'salt particles', and rises to 500% for carbonaceous particles. Thus, maritime aerosols attenuate radiation primarily by scattering, whilst carbonaceous particles attenuate radiation by absorption. Figure 3.5 shows the effect on solar radiation of a dust storm in the southern Sahara. It is noticed that the direct solar radiation is dramatically reduced by the dust, although the global solar radiation, i.e. the radiation from all directions, which includes the scattered radiation, is only reduced by approximately 15%. Hence, the major influence of the dust is the scattering of the light field rather than the absorption of radiation.

The last and most significant influence on solar radiation is that of clouds. Cloud droplets are typically larger than aerosol particles, usually varying between 1 and 20 μm in diameter. In this size range, the scattering is dominated by diffraction, which tends to produce scattering at small angles from the angle of incidence. This diffraction-dominated scattering does not tend to be dependent on wavelength. The absorption to back-scattering ratio for clouds is very small, and therefore the predominant influence on solar radiation is the back reflection. Particularly spectacular is the reflection from cumulus clouds on an otherwise clear day. Of course, ice is an important scatterer of radiation as well, but it tends to be more effective at scattering radiation at larger angles from the angle of incidence than water droplets, and therefore, close to the Sun, ice clouds appear less bright than water clouds. However, the opposite is the case for clouds in the direction away from the Sun. In addition, because the ice crystals are hexagonal, plate and columnar prism optical processes are important,

---

Figure 3.5 Solar radiation attenuation during a dust storm. (a) Normal incidence of radiation during the dust storm (13 November 1977) and clear sky (15 November 1977) as a function of local time. (b) Same as for (a) except for global radiation (direct + diffuse). Note the large depletion of direct solar radiation but the smaller depletion of global radiation and these produce a fascinating series of optical phenomena. The more common phenomena include the appearance of a 22° halo around the Sun and the production of parhelia, or mock Suns.

The quantitative influence of clouds on solar radiation is dependent on:

(i)   The solar elevation.
(ii) The thickness of the cloud.
(iii) The size, distribution and amount of liquid water in the cloud.

Though absorption is small, typically less than 10% and probably often as low as 5%, it is enhanced by scattering, which increases the path length within the cloud and hence the probability of water-vapour absorption. The transmission of the direct beam is also dependent on the type of cloud. For example, the direct beam is significantly attenuated by low-level clouds, and therefore a stratocumulus cloud of 400 m thickness will be sufficient to obliterate the Sun's disc. On the other hand, cirrus clouds allow the Sun's disc to be seen. As the thickness of the cloud increases, the transmission decreases, and the albedo increases and approaches a value between 80 and 90%. A decrease in the elevation of the Sun will also result in a decrease in the transmission of the direct beam and an increase in back-scattering by clouds. A generalized quantitative analysis is not possible, however, because of the differences between clouds of the same general type.

Figure 3.6 shows a detailed breakdown of the downward and upward components of solar radiation measured through low-level stratocumulus cloud, which is particularly common over the ocean. The cloud is 200–400 m thick on average and, in common with other low-level clouds, it has a high liquid water content of between 0.1 and 0.5 g m^{-3}. The albedo of the cloud is 68% and, of the solar radiation not reflected, 25% is transmitted through to the surface and 7% is absorbed by the cloud. Though the absorption is a relatively small contribution to the attenuation of the incident solar radiation when compared with the amount reflected, it is, in this case, equivalent to 52 W m^{-2}. This is sufficient to warm the upper part of the cloud by 20–40 K day^{-1} and offset the long-wave cooling from the cloud top. It is noted that only a small fraction of the downward solar radiation is reflected from the surface and this is consistent with the general observation of the low albedo over ocean surfaces.

Figure 3.6   The observed short-wave (S) fluxes measured from three aircraft during the JASIN experiment (August 1978), corrected to a solar zenith angle of 43.7°, for a stratocumulus cloud. The calculated heating-rate profile is shown in the right-hand diagram

For a non-uniform broken cloud, it is more difficult to characterize and quantify the solar radiation. The solar radiation may vary by 50% of the average value between clear and cloudy air. In addition, side reflections from the cloud may enhance the radiation above that measured under clear-sky conditions. For these reasons, it is necessary to obtain statistical relationships between the fraction of cloud cover and the solar transmission. In this respect, satellite techniques can be particularly powerful in estimating albedos for ensembles of clouds.

3.3   Absorption of planetary radiation

The predominant absorption of planetary radiation is by water vapour, carbon dioxide and ozone. All of these gases consist of triatomic molecules which have vibration–rotation absorption bands in the infra-red region of the spectrum, unlike diatomic molecules, such as oxygen and nitrogen, and the noble gases which exist as single atoms.

The detailed properties of the absorption curves are complicated, particularly for water vapour, as illustrated by Figure 3.2, but some bands are much more intense than others. These strong bands occur in the near infra-red at 1.4, 1.9, 2.7 and 6.3 μm for water vapour, and the continuum absorption band stretches from 13 to 1000 μm. The importance of this latter absorption band cannot be overemphasized, since more than 50% of the planetary radiation occurs at wavelengths greater than 13 μm. Carbon dioxide has its principal absorption bands at 2.7 and 4.3 μm, with a further strong band at 14.7 μm. Ozone has bands at 4.7, 9.6 and 14.1 μm. The 9.6 μm absorption band coincides with the maximum of the planetary radiation emitted from the surface. One of the broad features of the absorption curve typical of the atmosphere as a whole is the absence of absorption between 8 and 13 μm, except for the 9.6 μm absorption of ozone. It is noted that this region between 8 and 13 μm is within the peak emission region of the planetary radiation spectrum, and some 32% of the total planetary emission resides within it. The region is known as the 'atmospheric window' because, in the absence of clouds, planetary radiation should pass through the atmosphere almost unimpeded. However, recent work has shown that, with high water-vapour concentrations typical of tropical latitudes, additional absorption peaks appear in the window region as the result of the formation of water-vapour dimer, $(H_2O)_2$. The water dimer is formed by the joining of two water molecules by hydrogen bonding. Although it exists only in low concentrations when compared with water vapour, it has significant rotation–vibration absorption bands in the atmospheric window. With very high water-vapour concentrations, typically greater than 20 mb in tropical environments, the dimer has an important influence in reducing the long-wave radiation lost to space. The presence of this dimer may lead to doubling of the absorption at 10 μm (under conditions favourable to its formation). Other minor gases, such as nitrogen oxides, methane and carbon monoxide, also have absorption peaks in the infra-red spectrum, and therefore contribute to the absorption of planetary radiation.

In addition to gaseous absorption, the continuous absorption throughout the infra-red spectrum of liquid water droplets is also of prime importance in the absorption of planetary radiation. At a wavelength of 10 μm, a liquid water film only 60 μm thick is sufficient to absorb 95% of the incident radiation, although care has to be exerted to distinguish between the absorption by plane water surfaces and by spherical water droplets. In the latter case, scattering becomes an important process. As an example, a cloud 200 m thick with a liquid water concentration of $0.1\,g\,m^{-3}$ would absorb about 90% of the incident planetary radiation. Therefore, a reasonable thickness of unbroken cloud would be sufficient to absorb all the incident planetary radiation. Ice has a similar infra-red absorption to liquid water, but the total water content in ice clouds is generally much less than that found in mixed and water droplet clouds lower in the troposphere because of the low temperatures which prevail. Hence, ice clouds will not be as effective in absorbing infra-red radiation. Typically, cirrus cloud has an infra-red absorptivity of 35%, compared with an absorptivity of 100% for stratocumulus.

Additional absorption by aerosols occurs, but this effect is not as marked in the infra-red as it is at visible wavelengths because of the λ^{-1} dependence. However, the importance of hygroscopic nuclei when considering infra-red absorption cannot be overstressed, since the particles will grow by condensation and absorption will increase because of the water which becomes attached to the nuclei. It should therefore be expected that a significant increase in absorption of planetary radiation will occur as the humidity rises, and such an increase could well be anticipated in the marine atmosphere and over industrial areas where hygroscopic nuclei may be produced in large concentrations.

3.4 Vertical temperature profile and its relation to radiation

The mean vertical temperature profile for the atmosphere is shown in Figure 3.7. The most important features are the three turning points, known as the tropopause, the stratopause and the mesopause, which allow a quantitative division of the atmosphere into four distinct layers; the troposphere, the stratosphere, the mesosphere and the thermosphere. The troposphere accounts for 80% of the mass of the atmosphere and it is a region of strong vertical mixing. It is associated with our experience of 'weather' and is most important in the condensation of water vapour, the average decrease of temperature with height being approximately $6.5\,K\,km^{-1}$. The stratosphere contains about 19.9% of the mass of the atmosphere and it is a

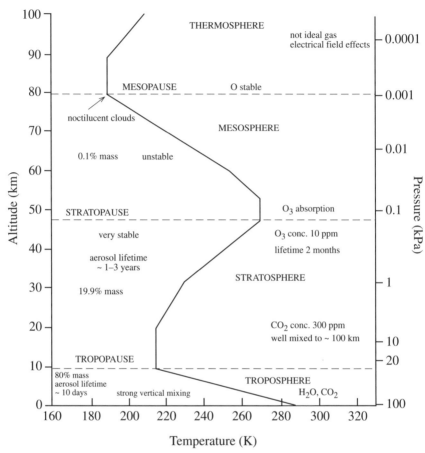

Figure 3.7   Global vertical temperature profile, delineating the four major regions of the atmosphere: the troposphere, stratosphere, mesosphere and thermosphere

region of very pronounced stability. The temperature increases with height at a rate of approximately $2\,K\,km^{-1}$, compared with a *decrease* of $9.8\,K\,km^{-1}$ expected for a dry neutrally buoyant atmosphere. The stability of the stratosphere is most marked between 30 and 50 km, where the absorption of solar radiation by ozone reaches a maximum. It is noted that the maximum ozone concentration occurs in the lower stratosphere. Although the tropopause forms a pronounced stability barrier between the stratosphere and the troposphere, it is highly variable in position. Occasionally, the tropopause may extend down to 6 km and, in these conditions, stratospheric air may be drawn into the troposphere, most commonly along frontal boundaries. There is a high ozone concentration in this stratospheric air and it is therefore possible to trace, from ozone concentrations, its incursion into the troposphere. The inherent stability of the stratosphere inhibits the vertical mixing of aerosols, and therefore they have a long residence time, typically of the order of 1–3 years, compared with some 10 days in the troposphere. This fact has important consequences for the climatic effects of volcanic aerosols. The El Chichon explosion of 1982, though of smaller magnitude than the Mount St Helens eruption of 1980, injected considerably more volcanic aerosol into the stratosphere. The dust cloud, composed mainly of sulphur particles and sulphuric acid droplets, was tracked across the Pacific Ocean for many months following the explosion and it produced significant reductions in solar radiation in Hawaii, over 6000 km from the volcano in Mexico, as has previously been stated. The Pinatubo eruption in June 1991 injected about three times as much sulphur dioxide into the stratosphere, as did the El Chichon eruption. This resulted in a global surface cooling of 0.5°C about 12 months after the explosion. Man-made pollutants will have similar long residence times in the stratosphere and it is in this region, therefore, that their climatic effects may be first observed. Of particular concern has been the input of exhaust gases from high-flying, supersonic aircraft, which may initiate photochemical reactions resulting in the reduction of the ozone concentration in the stratosphere and a subsequent increase in the amount of ultraviolet radiation incident on the Earth's surface.

The mesosphere contains 0.1% of the mass of the atmosphere and it is less stable than the stratosphere, with temperatures decreasing with height to reach a minimum value at 80 km. The temperatures are sometimes sufficiently low to produce freezing of even the low concentration of water vapour present and to produce noctilucent clouds. These clouds are normally translucent to the direct solar radiation incident upon them, but they can sometimes be seen in summer. Above the mesopause is the thermosphere, where the temperature increases steadily with height. In this region, the molecules become very well separated and collisions become so infrequent that we can no longer treat the atmosphere as an ideal gas. Furthermore, the thermosphere does not behave as a simple fluid because electrically charged particles will move differently to neutrally charged particles.

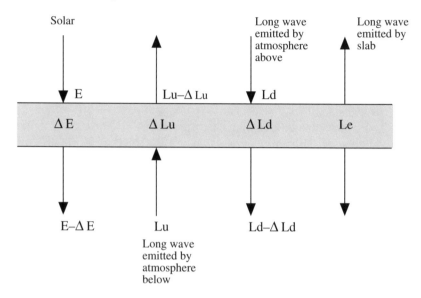

Figure 3.8 Radiation budget of a thin layer of the atmosphere (see text for nomenclature)

Consideration of the actual processes which produce the observed temperature distribution in the lower 100 km of the atmosphere is now in order. The detailed processes of absorption of solar radiation and long-wave planetary radiation in the atmosphere have already been discussed, and this information may be used to obtain a model of the vertical distribution of radiation and temperature. Consider a horizontal slab of the atmosphere, as depicted in Figure 3.8. The net rate of heating (or cooling) of the slab will depend on:

(i)   The absorption of solar radiation, ΔE.
(ii) The absorption of long-wave radiation from the atmosphere above, ΔL_d, and below the slab, ΔL_u.
(iii) The re-emission of long-wave radiation at the slab temperature, L_e.

Hence, following the above notation, the energy budget of the slab is given by:

$$mC_p \frac{\mathrm{d}T}{\mathrm{d}t} = \Delta E + \Delta L_u + \Delta L_d - 2L_e \qquad (3.1)$$

where m is the mass of the slab and C_p is the specific heat of air at constant pressure.

In radiative equilibrium, $dT/dt = 0$, i.e. the rate of change of temperature with time is zero, and therefore the radiation absorption is balanced by the radiation re-emission. As has been noted earlier, the absorption in a clear, cloud-free atmosphere depends on:

(i) The mass of absorbing gas.
(ii) The path length through the gas.
(iii) The absorption coefficient for the gas for each wavelength.

Ideally, the absorption coefficients would be calculated for each radiation wavelength between 0.1 and 100 μm, but this is very time consuming, even with high-speed computers, and therefore approximations to the absorption spectra are made. The simplest approximation is to associate a single absorption coefficient with the solar radiation band and another with the long-wave planetary radiation band. Similarly, for emission, an effective emissivity, ϵ, can be defined, which depends on the mass of carbon dioxide, water vapour and ozone. Hence, if the vertical distributions of these gases are known, then the atmosphere can be divided up into a number of levels, and equations similar to equation 3.1 may be written down for each level. The long-wave emission term, L_e, where $L_e = \epsilon \sigma T^4$, can be adjusted by iteration until it balances the other terms in the equation. Figure 3.9 shows the results of this type of calculation for a cloud-free atmosphere composed of carbon dioxide, water vapour and ozone. The results of the same calculation for (i) carbon dioxide only, (ii) carbon dioxide and water vapour only, and (iii) water vapour only, are also shown in Figure 3.9. It can be seen clearly that, without ozone absorption, the temperature would decrease with height up to 100 km and the stable stratosphere would not exist. This profile is similar to the atmospheric profiles of Venus and Mars, where ozone occurs in insufficient quantities to produce a stratosphere. Another interesting feature is the role of water vapour, which is very important in the absorption in the lower troposphere. In a water-free atmosphere, it is clear that the troposphere and the surface temperature would be considerably cooler than present observations show. It is interesting to note the shielding effect of tropospheric water vapour on the stratospheric temperature. In the absence of water vapour, long-wave radiation is transmitted from the surface into the stratosphere, and hence produces an increase in

Figure 3.9 Radiative equilibrium temperature profile for various atmospheric absorbers in the absence of clouds. The absorbers are water vapour (H_2O), carbon dioxide (CO_2) and ozone (O_3). The effects of both long-wave radiation and solar radiation are included

stratospheric temperature. The carbon dioxide does contribute significantly to the warming in the atmosphere but, because it is well mixed throughout the atmosphere, it tends to warm the stratosphere as much as the troposphere. It is noted that in the realistic model, as illustrated by Figure 3.9, the calculated surface temperature is 330 K, whilst the observed surface temperature is approximately 288 K. The reason for this discrepancy is that the vertical temperature gradient in the lowest 10 km is unstable (i.e. $\partial\theta/\partial z < 0$) for the radiative equilibrium profile. If the model is adjusted to a stable temperature profile ($\partial\theta/\partial z = 0$), then the surface temperature approaches the observed value and the temperature in the upper troposphere is increased. The assumption inherent in this temperature profile stabilization is that convection will continually readjust the vertical temperature profile by transferring heat from the surface to the upper troposphere. Hence, the radiation will always be endeavouring to produce an unstable gradient near the surface but, because of the efficiency of convection in the vertical transfer of heat, the surface will not come into radiative equilibrium. However, very close to the surface, below a height of about 1 m, unstable temperature gradients may be maintained. This occurs when the solar heating of the surface is large and small-scale convection near the surface is not sufficiently vigorous to completely adjust the vertical temperature gradient.

Such model results will be complicated by clouds, aerosols and horizontal variations in the distribution of water vapour and ozone in the atmosphere. It is model calculations such as these, however, which have been used to obtain estimates of the increase in surface temperature as the result of increasing levels of carbon dioxide in the atmosphere.

The fluxes of radiation for the observed global atmosphere are indicated in Figure 3.10, which assumes a solar flux of 100 units at the top of the atmosphere but, because of the variability of the absorption and reflection processes, the diagram can act only as a guide to the magnitude of the processes involved. In common with the approach in Sections 3.2 and 3.3, the solar and planetary radiation will be divided into two separate streams. Of the 100 units of solar radiation incident on the top of the atmosphere, about 50 units are absorbed by the Earth's surface, 30 units are reflected back to space and 20 units are absorbed by the atmosphere. The major reflector of solar radiation is the clouds, whilst the main absorbers are the atmospheric aerosol, water vapour and ozone. It is noted that, globally, the Earth's surface reflects only four units back to space, on account of the low albedo of the oceans. The Earth's surface and atmosphere, to maintain thermal equilibrium, must lose heat at the same rate as it is absorbed from solar radiation. The planetary radiation budget for the surface and atmosphere is rather more complicated than for solar radiation because of the re-emission of radiation by the atmosphere, known as the 'greenhouse effect'. For instance, 115 units of long-wave radiation are emitted from the Earth's surface, of which 109 units are absorbed by the atmosphere and the remaining six units are lost to space, through the atmospheric window. The atmosphere re-emits 95 units back to the Earth's surface, and therefore the long-wave flux lost from the surface is only 20 units. Furthermore, the surface gains 50 units by solar radiation, and therefore has a net surplus of 30 units of total radiation (solar + planetary). To maintain the thermal equilibrium of the surface, it is necessary to transfer this surplus heat to the atmosphere by convection and by the evaporation of water. Globally, about 75% of

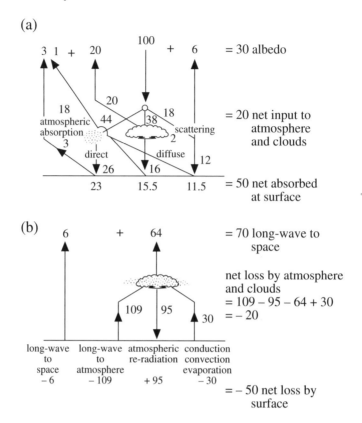

Figure 3.10 The globally and annually averaged components of the solar (a) and long-wave (b) radiation streams. The proportion of the radiation intercepted by various components of the atmosphere and the Earth's surface is normalized to the incident radiation at the top of the atmosphere (100)

the surplus heat is used to evaporate water, though this fraction will vary between land and ocean. Over land about 50% of the energy is used for evaporation, whilst over the ocean about 90% of the available energy is used. The atmosphere absorbs 109 units of long-wave radiation emitted from the surface, but re-emits 95 back to the surface and 64 to space. Therefore, the atmosphere has a net deficit of long-wave radiation of 50 units. Furthermore, solar absorption by the atmosphere accounts for only 20 units, and therefore there is a net total radiation deficit of 30 units for the atmosphere. The global atmosphere will always be continually cooling by radiation processes, at the rate of 30% of the incident solar radiation $\sim 100\,\mathrm{W\,m^{-2}}$. The mass of the atmosphere per unit area is $\sim 10^4\,\mathrm{kg\,m^{-2}}$ and the specific heat at constant pressure is $1000\,\mathrm{J\,kg^{-1}\,K^{-1}}$, and therefore the cooling rate is $10^2/(10^3 \times 10^4)\,\mathrm{K\,s^{-1}}$ or $1\,\mathrm{K\,day^{-1}}$. Hence, the atmosphere will continually cool by radiation at $\sim 1\,\mathrm{K\,day^{-1}}$. Therefore, to maintain the thermal equilibrium of the atmosphere, it is necessary to transfer heat from the Earth's surface by convection and by the release of latent heat of condensation.

Figure 3.11 shows the net heating rates observed for the atmosphere. The continual convection of heat, both horizontally and vertically,

Figure 3.11 (a) Mean total radiative heating $(\mathrm{K\,day^{-1}})$ for December–February. (b) Mean total radiative heating $(\mathrm{K\,day^{-1}})$ for June–August. Areas of heating are shaded

implies that the atmosphere will never be in radiative equilibrium. It is noted that, in general, the tropical stratosphere is always heating up, mainly as the result of ozone and carbon dioxide absorption, whilst the troposphere is cooling at rates between 0.5 and $1.8\,\mathrm{K}\,\mathrm{day}^{-1}$, except over the poles where some heating is observed in the summer hemisphere. The largest cooling rates occur over the surface and at the top of the equatorial troposphere, where high temperatures and high humidities result in a large loss of long-wave radiation. To maintain this cooling rate, heat must be continually convected from the surface into the atmosphere.

3.5   The absorption of solar radiation in the ocean

The ocean absorbs the largest fraction of the solar radiation incident on the Earth's surface because of its great area and its low albedo. The manner in which this radiation is absorbed is of importance in understanding both the heating distribution in the ocean and the stability of the ocean. Furthermore, the biological activity of the ocean is primarily dependent on the direct absorption of the incident solar radiation by microscopic phytoplankton. The rate of phytoplankton growth, or primary production, is a major control on the development of food chains which ultimately determine the productivity of the fisheries and the population of the larger sea mammals at the ends of these chains.

It can be shown that attenuation of solar radiation by pure sea water can be represented by an exponential law of the following form:

$$E_\lambda(z) = E_{\lambda 0} \exp(-k_\lambda z) \tag{3.2}$$

where k_λ is the attenuation coefficient at wavelength λ and $E_\lambda(z)$ is the solar radiation intercepted at a depth z below the sea surface. $E_{\lambda 0}$ represents the solar radiation incident on the sea surface at a wavelength λ. The variation of k_λ with wavelength is depicted in Figure 3.12a for the solar spectrum. The main feature is that the coefficient varies by a factor of 10^7, with a minimum value at $0.46\,\mu\mathrm{m}$, in the blue region of the visible spectrum, and maximum values in the ultraviolet and near infra-red regions in the spectrum. The inverse of the absorption coefficient, $1/k_\lambda$, gives a measure of the depth of penetration of the radiation, $z_p(\lambda)$. Substituting in equation 3.2:

$$E_\lambda(z_p) = E_{\lambda 0}\ e^{-1} \text{ or } 0.37E_{\lambda 0}$$

Thus, the penetration depth, $z_p(\lambda)$, is the depth to which 37% of the incident radiation penetrates. At 0.46 μm, this depth is approximately 55 m, whilst at 0.3 μm, in the ultraviolet, it is 10 cm and at 1 μm, in the infra-red, it is 1 cm. The strong absorption of ultraviolet radiation by sea water was important in shielding early organisms from early doses of ultraviolet radiation at a time when the ozone layer in the atmosphere was non-existent, as previously discussed in Section 2.1. Less than 1% of the most penetrating blue radiation will reach 500 m and even the human eye, which has a truly remarkable ability to see objects at radiation levels as low as $50 \times 10^{-6}\,\mathrm{W\,m^{-2}}$, would not be able to detect any solar radiation below a depth of 800 m. For these reasons, the majority of the ocean's depths are dark and do not benefit from solar radiation absorption.

The absorption of the solar spectrum in pure sea water, depicted in Figure 3.12b, shows that the majority of infra-red radiation is absorbed in the first 10 cm of the ocean and that all of it is absorbed within 1 m of the surface. Thus, over 55% of the incident solar radiation is absorbed in the first metre of sea water but, once the infra-red and ultraviolet components have been eliminated, the rate of absorption with depth decreases, progressively removing the red and orange components of the solar spectrum and leaving only the most penetrating blue component. The sea water, therefore, acts as a monochrometer, removing all colours except blue below a depth of 10 m. The blue colour of deep sea water is predominantly the result of the absorption process, rather than Rayleigh scattering from sea-water molecules. A person looking downwards into the ocean sees light which has been back-scattered from successive layers of water. Since all of the other colours except blue have been absorbed very rapidly, a large proportion of the back-scattered light comes from layers where only blue light has penetrated. The blue colour of the ocean surface is also enhanced by the reflection of 'sky light'.

Thus far, only the absorption of solar radiation by pure sea water has been considered. Only in regions of very low biological productivity situated at a considerable distance from estuarine environments does the pure sea-water absorption approach reality. The centres of the large

Figure 3.12 (a) Absorption of radiation by sea water. (b) The complete spectrum of downward solar irradiance in the sea between the surface and 100 m

sub-tropical gyres, such as the Sargasso Sea, are such regions of very clear ocean water. In regions of significant primary productivity, the colour of the sea water changes from dark blue to blue-green. The minimum absorption no longer occurs at 0.46 μm, but it is shifted towards the green and occurs at approximately 0.48 μm. The shift is primarily the result of the presence of chlorophyll which absorbs strongly in the purple and blue, and less significantly in the red part of the spectrum. In addition to the colour change, the presence of chlorophyll also reduces the overall penetration depth of solar radiation. In very productive waters, such as the equatorial upwelling region, the penetration depth for blue light at 0.46 μm is $\sim$10 m, compared with a penetration depth in pure sea water of 55 m. However, the most significant effects are found in coastal waters over the continental shelves. Here, strong tidal mixing and large inputs of suspended solids from rivers increase the attenuation coefficient dramatically and penetration depths of $\sim$1 m for blue light are observed. In addition, chlorophyll pigment and other biologically-derived pigments produce variations in the perceived colour of sea water. For example, a yellow pigment called gelbstoft, which is prevalent in European coastal waters, has a very strong absorption in the blue part of the spectrum, and thus the most penetrating wavelengths occur around 0.55 μm in the yellow part of the visible spectrum. The presence of large quantities of suspended mineral sediment in some coastal waters tend to impart a grey colour because of scattering by the mineral particles. Occasionally, a red pigment, produced by a particular species of phytoplankton, can cause a phenomenon known as 'the red tide'. This pigment can be remarkably toxic to many other organisms, and it is therefore usually associated with a temporary absence of fish and other forms of marine life. Red tides are observed, for example, in Southampton Water in the United Kingdom, usually after a period of warm weather.

Table 3.1 shows the typical penetration depths of solar radiation for different types of water masses. The most important point to notice is that, because of the presence of biologically-derived pigments, and mineral and calcareous particles in the ocean, the average penetration depth is much less than that of pure sea water. All of the incident solar radiation is effectively absorbed within the uppermost 50 m of the ocean. Thus, the ocean is warmed from above and the formation of a warm surface layer enhances the stability of the ocean, in the same way

Table 3.1   Penetration depth (metres) of solar radiation (0.3--2.4 μm) in typical ocean and coastal water masses

Water type	Example of water type	Depth for attenuation of 90% of incident solar radiation	Depth for attenuation of 99% of incident solar radiation
Ocean I	Sargasso Sea	33	85
Ocean II	North-eastern Atlantic	14	45
Ocean III	Equatorial East Pacific (upwelling)	8	25
Coastal 1	Continental shelf-break	7	20
Coastal 9	Baltic Sea	2	5

that the stability of the stratosphere is enhanced by ozone absorption and, in marked contrast to the troposphere, where radiative heating of the surface maintains instability and convection.

3.6   Diurnal and seasonal temperature cycles in the ocean

We will now consider the temperature in the surface layers of the ocean in response to the diurnal and seasonal cycles of solar radiation. Figure 3.13 shows the theoretical heating rates in the ocean in the absence of mixing. The temperature changes, in the upper 5 cm of the ocean, are as large as $40\,\text{K}\,\text{day}^{-1}$ during summer. This is because of the rapid absorption of the near infra-red solar radiation shown in Figure 3.12.

The observed temperature changes over a diurnal cycle are, however, considerably smaller, being typically in the range of 1–2 K over a day. The vertical mixing caused by surface wind stress and waves during the day mix the warm surface layers with underlying cooler layers. Even on a perfectly calm and clear day, there is a cooling of the surface 1 mm of the ocean, known as the surface skin, caused by evaporation and long-wave radiation from the ocean surface to the overlying atmosphere. On calm days, the maximum warming occurs in the afternoon, about two hours after the maximum solar heating. The diurnal thermocline is destroyed during the night due to surface cooling, which produces vertical convection and mixing in this layer.

The seasonal temperature changes in the surface layer shown in Figure 3.14 are similarly governed by the two processes of the absorption of

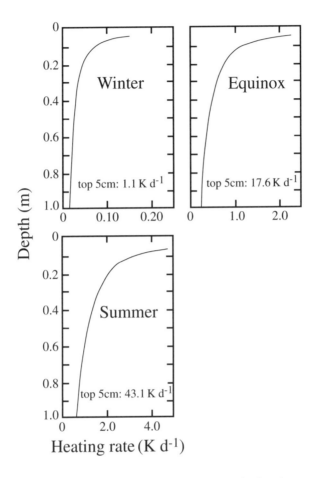

Figure 3.13   The theoretical daily temperature rise, in the absence of mixing, in the top metre of the ocean for clear sea water in winter, spring and summer at 60° latitude

solar radiation and vertical mixing. In late winter, the surface water column is homogeneous down to depths of one or two hundred metres, due to a combination of winter cooling and mixing. Warming of the surface layers starts to occur as the solar radiation increases in spring and a seasonal thermocline becomes established at the boundary of the warm surface layer. As more heat is absorbed, the buoyancy increases and the layer becomes more stable, and there is less vertical mixing. Once the seasonal thermocline is established, it is remarkably resilient to vertical mixing by transient storms. The surface layer

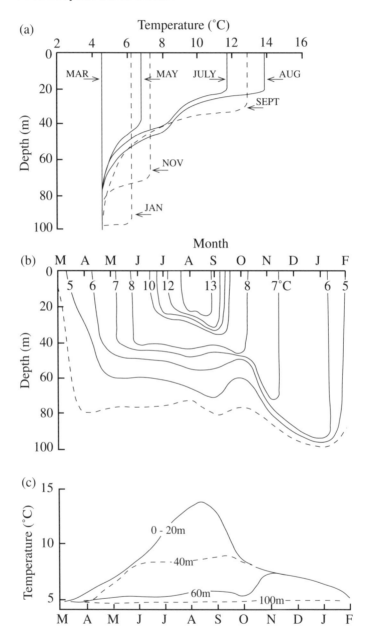

Figure 3.14 The annual cycle of temperature in the upper layers of ocean in the eastern North Pacific. The seasonal thermocline is the region of maximum vertical temperature gradient which occurs at depths of 30 m in summer and 100 m in winter

continues to increase in temperature for about two months after the summer solstice when the maximum solar radiation occurs. As the solar radiation decreases to the winter solstice, surface cooling by evaporation, sensible heat and long-wave radiation exceeds the solar heating, which results in surface cooling. This cooling, in conjunction with increased storminess, increases the vertical mixing of the warm surface layer, which in turn increases the cooling of the mixed layer. The heat added into the surface layer during the summer is removed during the autumn and winter, and warms the overlying atmosphere. As the heat is depleted, vertical mixing is able to penetrate to deeper levels in the ocean. In the North Atlantic, the maximum depth of surface mixing is typically between 100 and 200 m, though in some localities west of Scotland the mixing may extend to depths of 500 m or more. In the North Pacific, the presence of a halocline limits the depth of mixing to about 100 m.

4

Water in the Atmosphere

4.1 Introduction

The atmosphere is the smallest reservoir of water on the planet Earth, contributing only 0.001% of the total mass of water present. In comparison, the ocean contains 97%, the ice caps of Antarctica and Greenland contain 2.4% and the freshwater reservoirs contribute the remaining percentage. However, the effect of the water present in the atmosphere is by no means small. It has already been shown that water vapour in the atmosphere is an important absorber of both solar and long-wave planetary radiation, and that, upon condensation, liquid water drops can both reflect solar radiation back into space and intercept long-wave radiation from the surface and the atmosphere.

When water vapour condenses there is a large release of latent heat. For each kilogram of condensed water produced, 2.4 MJ of energy are released. This input of energy into the atmosphere is equivalent to slightly less than one-quarter of the solar flux incident on the Earth, or approximately $80\,\mathrm{W\,m^{-2}}$ on average, and it is, therefore, a major source of energy for driving the circulation of the atmosphere. For example, the trade-wind circulations of both hemispheres are enhanced by the release of latent heat in the inter-tropical convergence zones (ITCZ) which straddle the thermal equator. The latent heat of condensation of water vapour may be released many thousands of kilometres away from where the evaporation of the water molecule took place and the wind circulation in the lower troposphere is important in determining the location of the energy release. Globally, the precipitation rate must balance the evaporation rate, but relatively small variations in the wind circulation can drastically influence the spatial distribution of precipitation and produce devastating effects on agriculturally-based economies.

4.2   The moist atmosphere

Figure 4.1 shows the phase diagram for water within the range of atmospheric temperatures. The curve AB is the saturated vapour pressure (SVP) over a plane ice surface and the curve BC corresponds to the SVP over a plane liquid water surface. The point B represents the point where water vapour, liquid water and ice can co-exist in equilibrium. It is known as the triple point. Let us first consider the range of vapour pressure which can exist in the atmosphere. At temperatures of $-50°C$, typical of winter-time in the polar regions, the SVP over ice is only 4 Pa, whilst at 30°C, typical of the equatorial region, the SVP approaches 4300 Pa. Between 0 and 30°C, the SVP approximately doubles for each 10°C rise in temperature. Though there is a larger range of water-vapour concentration in the atmosphere, the maximum value is only 4.3% of the surface pressure.

In the atmosphere, liquid water can exist in a supercooled state. The SVP over a supercooled water surface is higher than the SVP over an ice surface, thus, a vapour in equilibrium with a supercooled water surface would be supersaturated with respect to an ice surface and water vapour would condense preferentially onto the ice surface (Figure 4.9). This is an important process in the formation of precipitation in cold clouds, which will be discussed in Section 4.5.

Water vapour is a gas with a molecular weight of 18 a.m.u., compared with a mean molecular weight of 29 a.m.u. for dry air, and hence a mixture of water vapour and dry air will have a lower density than an equivalent volume of dry air, assuming that temperature and pressure remain constant. The equation of state for moist air therefore requires another variable, namely the mass of water vapour, to determine the relationship between pressure, temperature and density. The equation of state for moist air gives the mean molecular weight, $\bar{m}$, of a mixture of dry air of mass M_d and of water vapour of mass M_v in a given volume as:

$$\frac{1}{\bar{m}} = \frac{M_d}{m_d(M_d + M_v)} \left[1 + \frac{M_v/M_d}{m_v/m_d} \right] \qquad (4.1)$$

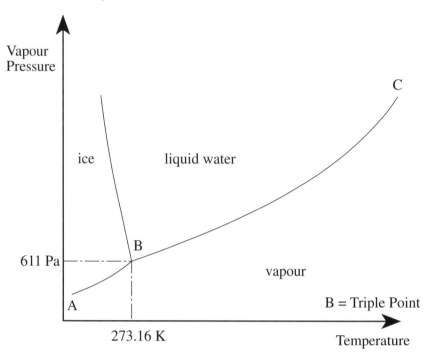

Figure 4.1   Phase diagram for water in the vapour, liquid and solid phases showing the triple point

where m_d and m_v are the molecular weights of dry air and water vapour, respectively.

The concentration of water in the atmosphere is defined by a mixing ratio, q, where

$$q = M_v/M_d \tag{4.2}$$

The mixing ratio is dimensionless, but meteorologists often express it in terms of grams of water vapour per kilogram of dry air. Since, for a given volume, masses are proportional to densities, the above equation may be rewritten as:

$$q = \rho_v/\rho_d \tag{4.3}$$

where ρ_v and ρ_d are the densities of water vapour and dry air, respectively. The ideal gas equations for dry air and water vapour yield the following:

$$\rho_v = \frac{e}{(R^*/m_v)T} \tag{4.4}$$

and

$$\rho_d = \frac{p - e}{(R^*/m_d)T} \tag{4.5}$$

where p is the total pressure of the dry air and vapour, and e is the vapour pressure of water. Water vapour is not in general an ideal gas but, at low concentrations, such as found in the atmosphere, the approximation is a valid one.

Substituting equations 4.4 and 4.5 into 4.3 yields:

$$q = \frac{e}{p - e}\left(\frac{m_v}{m_d}\right) \tag{4.6}$$

An approximate expression for q can be obtained by noting that the total pressure of the dry air and the water vapour is usually two orders of magnitude larger than the vapour pressure of water, and hence:

$$q \simeq \frac{e}{p}\left(\frac{m_v}{m_d}\right) \tag{4.7}$$

Substituting for M_v/M_d in equation 4.1 gives:

$$\frac{1}{\bar{m}} = \frac{1}{m_d}\left[\frac{1 + q(m_d/m_v)}{1 + q}\right] \tag{4.8}$$

Substituting equation 4.8 into the equation for a mixture of ideal gases, the ideal gas equation for a moist atmosphere is obtained, namely:

$$p\alpha = \frac{R^*}{m_d}\left[\frac{1 + q(m_d/m_v)}{1 + q}\right]T \tag{4.9}$$

This equation is similar in form to the ideal gas equation for dry air except for the addition of the factor involving the mixing ratio, q. An example of the use of this equation is now appropriate. If the vapour pressure of water in the atmosphere is 1000 Pa and the total pressure is 10^5 Pa, then q is found to be 6.22×10^{-3} or 6.2 g of water vapour per kilogram of dry air, and hence:

$$p\alpha = \frac{R^*}{m_{\mathrm{d}}}\left[1 + 3.78 \times 10^{-3}\right]T$$

At this stage it is useful to define a virtual temperature, T_{v}, where

$$T_{\mathrm{v}} = T\left[\frac{1 + q(m_{\mathrm{d}}/m_{\mathrm{v}})}{1 + q}\right] \qquad (4.10)$$

This virtual temperature is always higher than the actual temperature but, in practice, the difference is less than 5 K. In the middle latitudes the difference is typically 1 K. In physical terms, the virtual temperature is the temperature a volume of dry air would have if it had the same density as an equivalent volume of moist air and, because water vapour is a lighter gas, this temperature will always be larger than the actual temperature.

Another useful parameter is the relative humidity, r, given by:

$$r = \frac{q(T)}{q_s(T)}\,100\%$$

where q is the actual mixing ratio of the sample of air and q_s is the saturation mixing ratio at the same temperature. It gives a useful measure of the humidity of moist air mass relative to its saturation value, and hence 100% relative humidity corresponds to a saturated air mass.

4.3   Measurement and observation of water vapour

Direct measurements of the vapour pressure and the mixing ratio, q, though feasible in a laboratory, are not practicable for the routine measurements in the atmosphere. The most direct method is to pass air over a flat silver plate which is cooled, and the temperature at which a dew first forms is noted, i.e. the dew-point temperature, T_{d}, is measured. The water-vapour content of the air is unaltered during the cooling, as shown in Figure 4.2, and hence the saturated mixing ratio at the dew-point temperature is equal to the mixing ratio of the air at its original temperature, T. The saturated mixing ratio is a known function of dew-point temperature and it can, therefore, easily be obtained from tables. It is noted that the cooling process must take place at constant

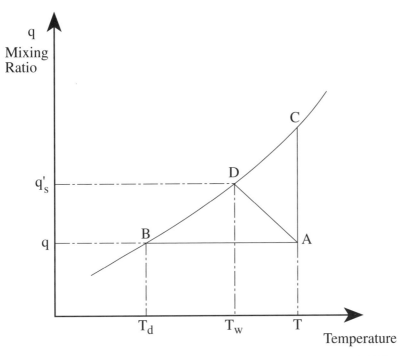

Figure 4.2 Dew-point temperature (T_d) and wet-bulb temperature (T_w) for a parcel of moist air at temperature T and mixing ratio q

atmospheric pressure to maintain a constant mixing ratio, as shown in equation 4.6. At lower temperatures, frost is more likely to form rather than dew and, in this case, the saturated mixing ratio with respect to ice must be used instead of that with respect to water.

A more practical method for surface measurements of humidity is that using the aspirated psychrometer. In this technique, air is drawn past two thermometers, one of which has its bulb covered with a muslin cloth which is kept moist with distilled water. The wet-bulb temperature always lies between the dry-bulb temperature and the dew-point temperature, except at saturation when all three temperatures coincide, as shown in Figure 4.2. The wet bulb evaporates water into the surrounding air, which:

(i) Decreases the temperature of the wet bulb due to the latent energy input required to produce evaporation.
(ii) Increases the mixing ratio of the surrounding air.

Hence, the saturated air surrounding the wet bulb can be represented by a point C in Figure 4.2, and the difference between the saturation mixing ratio at the wet-bulb temperature (q'_s) and the mixing ratio of the environmental air (q) can be shown to be proportional to the difference between the wet- and dry-bulb temperatures, i.e.

$$T - T_w = A[q'_s - q]$$

where A is a constant which depends on the ventilation rate of the instrument. q'_s can be obtained from tables, and hence q is obtained.

Other techniques include the measurement of the change in length of hair or skin with changing humidity. A human hair will expand by 2% in changing from an absolutely dry to a saturated environment, and this property can be used to measure relative humidity where cheap and reasonably reliable instrumentation is required. This technique is used in some types of radiosonde, but it cannot be used to measure low humidities, and the hair component tends to exhibit strong hysteresis after becoming saturated in clouds.

The distribution of water vapour in the atmosphere is shown in Figure 4.3. The mixing ratio reaches a maximum in the equatorial zone, and rapidly decreases with both altitude and latitude. This is a general reflection of the variation of the saturated mixing ratio with temperature. The majority of the water vapour occurs below 50 kPa, although the small amounts present in the upper troposphere and the stratosphere have already been shown to be important for the radiation balance in these regions. If all the water vapour were to be condensed out of the atmosphere, it would result in an average global precipitation of 2.5 cm, ranging from 4 cm in the equatorial regions to less than 0.5 cm over the poles. This can be compared with the observed global precipitation of 100 cm year^{-1}. The ratio of the water content of the atmosphere to the observed precipitation rate gives the time scale for the replacement of the atmospheric water vapour by evaporation from the surface. Thus, the average lifetime of a water molecule from evaporation at the surface to precipitation is approximately nine days, and this is sufficient for the water molecule to be carried around the globe by the tropospheric wind system before it is precipitated. The relative humidity distribution shown in Figure 4.3 gives an indication of the efficiency of the water-holding capacity of the atmosphere. In the equatorial zone, between 0 and 10º

Figure 4.3   (a) The longitude-averaged specific humidity (q) in northern winter (October–March). Values are in $g\,kg^{-1}$. (b) The longitude-averaged relative humidity (q/q_s) in northern winter. Values are percentages

and at latitudes greater than 50º, the surface layers of the atmosphere are close to saturation. However, in the sub-tropical zone of the mid-troposphere, the relative humidity is low, attaining values of only 30% as a result of the descent of air from the upper troposphere where it has a very low water content. This air will warm by adiabatic compression as it moves downwards through the troposphere, but will retain its low water content. Hence, the relative humidity will decrease as one moves from the upper to the lower troposphere, although close to the surface, evaporation into the air mass will act to increase the relative humidity.

4.4   Stability in a moist atmosphere

It has already been shown that water vapour will change the thermodynamic properties of a sample of air by:

(i)   Decreasing the mean molecular weight of the moist sample in comparison with a sample of dry air.
(ii) The release of the latent heat of condensation when precipitation from the moist sample occurs.

The change in the first property is directly proportional to the mass of water vapour in the given sample of air, whilst the second property is proportional to the mass of water vapour which condenses to form liquid water. In order to obtain a qualitative feel for the relative effects of the two processes, Figure 4.4 illustrates the temperature history of a parcel of moist air, typical of the middle latitudes, as it is lifted through the atmosphere. At the surface, it is assumed that the air has a water-vapour content of 3.5 g per kilogram of dry air, which corresponds to a virtual temperature 1 K higher than the actual temperature. From the surface to a height of 1 km, the parcel of air cools at the dry adiabatic lapse rate[†] of 9.8 K km^{-1}. At the point at which the condensation level is reached, release of latent heat slows down the rate of cooling of the parcel. The amount of latent heat released between 1 and 3 km depends solely on the difference between the saturation mixing ratios of the air at the two levels, which is approximately 7 g per kilogram of dry air in this case. This is equivalent to the release of 17 500 J for each kilogram

---

†A moist unsaturated sample of air will cool at the dry adiabatic lapse rate, to a good approximation.

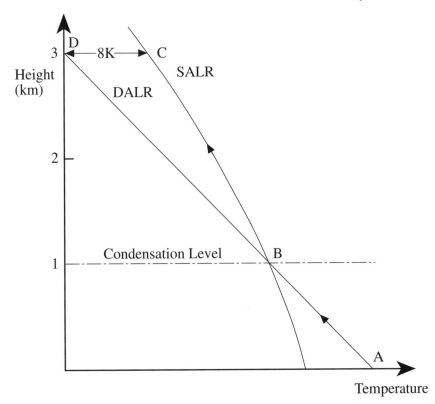

Figure 4.4 Typical ascent of a moist air parcel (*ABC*) and a dry air parcel (*ABD*). See text for details

of dry air. The temperature difference between a moist parcel of air at 3 km and a dry parcel of air at the same height is approximately 8 K, which is due solely to the release of the latent heat of condensation. It is noted that the difference between the actual and the virtual temperature is generally smaller, though not negligible, compared with the difference between dry air parcels and moist air parcels after condensation.

Now consider the change in temperature with height of a parcel of air at the level of condensation. If the parcel rises an infinitesimally small distance, dz, then the difference in the saturation mixing ratio, brought about by adiabatic cooling, will be $-dq_s$. The negative sign denotes a decrease in the saturation mixing ratio with increasing height. The latent heat released for each kilogram of dry air is given by $-L \, dq_s$, where L is the latent heat of condensation, and this heat is, therefore, an additional source of energy, equivalent to the term dQ in equation 2.5. Thus:

$$-L dq_s = C_p dT - \alpha\, dp \qquad (4.11)$$

and from the hydrostatic equation:

$$\alpha\, dp = -g\, dz \qquad (4.12)$$

hence, by substitution in equation 4.11:

$$-L\, dq_s = C_p\, dT + g\, dz \qquad (4.13)$$

Rearranging and dividing through by dz yields

$$\frac{dT}{dz} = -\frac{L\, dq_s}{C_p\, dz} - \frac{g}{C_p}$$

Since $dq_s/ds = (dq_s/dT)(dT/dz)$

$$\frac{dT}{dz} = \frac{-g}{C_p}\left[\frac{1}{1 + (L/C_p)(dq_s/dT)}\right] = \Gamma_s$$

where Γ_s is known as the saturated adiabatic lapse rate (SALR).

It is noted that as dq_s/dT is always positive, the SALR will always be less than the dry adiabatic lapse rate, given by $\Gamma_d = -g/C_p$. At a temperature of 20°C and at pressure of 10^5 Pa, the SALR is approximately $0.41\Gamma_d$ or $4\,\mathrm{K\,km^{-1}}$, whilst at 10°C it is $0.59\Gamma_d$ or $5.8\,\mathrm{K\,km^{-1}}$. At high levels in the troposphere, the saturation mixing ratio will become very small and the SALR will approach the dry adiabatic lapse rate asymptotically.

The addition of latent heat of condensation also increases the potential temperature of the air and hence, in any region of the atmosphere where condensation occurs, the potential temperature will no longer be conserved. The stability of the atmosphere will therefore be significantly changed by the condensation process. Consider a parcel of dry air displaced from its equilibrium position in a stable environment where the temperature gradient, $d\theta/dz$, is positive. As discussed in Section 2.6, the parcel will be subjected to a negative buoyancy force as it rises and this force will act to return it to the equilibrium position. Now consider a parcel of air which is saturated at the equilibrium position z_e. On rising a small distance δz, it will condense some of its water vapour and this will warm the parcel by an amount $\delta\theta_1$. If $\delta\theta_1 > \delta\theta$ (where $\delta\theta$ is the temperature change of the environmental air over a small distance δz), then the moist parcel will be warmer than its environment (path AB in Figure 4.5) and it will

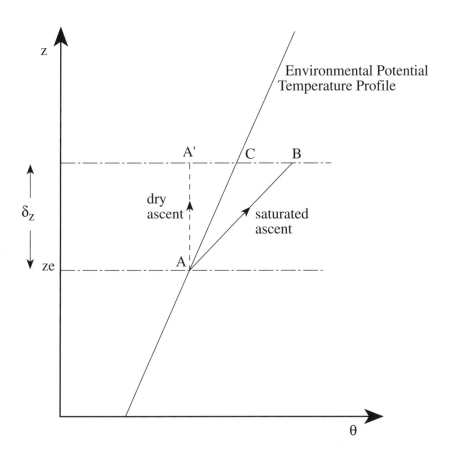

Figure 4.5 Conditional instability of a moist atmosphere. The parcel of air at A is in equilibrium with its environment. If the parcel is unsaturated and is perturbed upwards, it will follow AA′ and it will become cooler than its environment. Negative buoyancy forces will return it to its initial position and therefore the parcel is stable with respect to its environment. If the parcel is saturated, it will follow path AB and will become warmer than its environment. Buoyancy forces will accelerate the parcel upwards, away from its initial position, and it will now become unstable

accelerate upwards away from its equilibrium position. The result of the condensation process, therefore, is to make the previously stable environment profile unstable. Clearly, if the warming is insufficient, i.e. $\delta\theta_1 < \delta\theta$, the profile will still be stable. This form of instability is known as 'conditional instability' and is produced by the condensation process. However, it is not possible to determine the conditions for instability from the potential temperature alone and an equivalent potential temperature, θ_e, must be defined:

$$\theta_e = \theta \, \exp\left(\frac{Lq_s}{C_p T} \right)$$

This equivalent potential temperature is conserved in saturated conditions, and it is defined as the potential temperature a parcel of air would have if all the water vapour in the parcel were to be condensed and all the latent heat of condensation converted into internal energy. It has the property of remaining constant during a saturated adiabatic process. Thus, the stability condition for saturated air is determined from the vertical profile of the equivalent potential temperature rather than from the potential temperature.

If, in the case considered above, $d\theta_e/dz < 0$, then the vertical profile will be unstable for saturated air. Parcels of air would have positive buoyancy when they rose and they would accelerate away from their initial surroundings. A profile with $d\theta_e/dz > 0$ would be stable for both dry and saturated air. It is therefore possible to compile a simple stability table:

	$d\theta_e/dz < 0$	$d\theta_e/dz > 0$
$d\theta/dz < 0$	Always unstable	Not possible
$d\theta/dz > 0$	Conditionally unstable	Always stable

It is noted that the stability of an environmental temperature profile can be determined either from the equivalent potential and the potential temperature profiles, or by comparison of dry adiabatic and saturated adiabatic lapse rates.

An additional form of moist instability is termed the potential instability, and this occurs when the profile has a negative equivalent potential temperature gradient but the air is not saturated. In this case, the lifting of the air mass is required to cool the air to its condensation point and to release the instability. This type of instability is often associated with the advection of an upper-level, dry air mass over a very moist, low-level air mass. The advection produces a large negative equivalent potential temperature gradient. Usually, uplift by mountains or uplift ahead of a depression may be sufficient to release the instability, which will often be associated with vigorous convection and occasionally produce severe thunderstorms with hail.

It is noted that a vertical gradient of humidity may induce instability in unsaturated air. Consider a neutral temperature profile on to which is superimposed a humidity gradient. In this case the virtual temperature would be higher nearer to the surface than above and parcels of air would have positive buoyancy if they were displaced upwards, and would, therefore, accelerate upwards. This type of instability is probably most important in the surface boundary layer over the tropical ocean where the virtual temperature can be several degrees higher than the actual temperature. It is also possible to introduce instability by the evaporation of precipitation in unsaturated air. Precipitation falling from a cloud may evaporate and cool the air beneath to the wet-bulb temperature. The air will become negatively buoyant and accelerate downwards. This process is responsible for the sudden drop in temperature experienced in the downdraughts of thunderstorms and for the sudden gusts of wind, up to $20 \, \mathrm{m \, s^{-1}}$, felt at the surface. Re-evaporation of precipitation is particularly common in the tropics, where cloud bases are generally higher than those found in the middle latitudes.

4.5 Processes of precipitation and evaporation: the formation of clouds

In the previous discussion it was implicitly assumed that condensation would occur when the relative humidity reached 100%, i.e. when the vapour pressure was equal to the saturation vapour pressure. However, the saturation vapour pressure described in Section 4.2 was defined with respect to a plane liquid water surface and not a curved surface, such as a rain droplet. In a sample of pure moist air, where all particles have been removed, condensation will not occur at 100% relative humidity, but will only commence spontaneously at a relative humidity of 800%, equivalent to a supersaturation of 700%. The reason for this phenomenon is that the saturated vapour pressure over a curved surface is higher than over a plane surface. The relationship between the saturated vapour pressure over a curved surface, e_r, and that over a plane water surface, e_∞, is given by:

$$\ln \frac{e_r}{e_\infty} = \left(\frac{2\gamma m_w}{\rho_w R^* T} \right) \frac{1}{r} \qquad (4.14)$$

where γ, m_w and ρ_w are the surface tension, molecular weight of water and density of water, respectively, r is the radius of curvature and R^* is the universal gas content. At a temperature of 273 K, the term in the bracket is 1.19×10^{-9} m.

If the value $e_r/e_\infty = 8$ is substituted in equation 4.14, a value of $r \simeq 0.6 \times 10^{-9}$ m is obtained, and this is the typical size of a small group of water molecules. Therefore, condensation in pure air samples will occur on groups of water molecules randomly brought together by Brownian motion. However, if a sample of air is taken directly from the atmosphere, condensation will occur at much smaller supersaturations, typically less than 1%. This is because of the presence of particles in the atmosphere which provide surfaces for condensation. These particles are known as condensation nuclei. The size of these nuclei, estimated by substitution into equation 4.14, is approximately 0.1 μm for a supersaturation of 1%. Basically, there are two types of condensation nuclei. The first type are hygroscopic nuclei and they are usually ionic salts, such as sodium chloride and ammonium sulphate, which dissolve readily when condensation occurs. They have the property that they are able to initiate condensation at relative humidities as low as 80%. The second type of condensation nuclei are the hydrophobic particles, which are not soluble in water and which require supersaturation conditions for condensation to occur.

Equation 4.14 has to be modified for hygroscopic nuclei. Figure 4.6 shows the SVP curves for pure water droplets and droplets containing dissolved hygroscopic nuclei. It is noted that the difference between the two SVP curves is most marked for small droplets having radii less than 0.1 μm. For larger water droplets, the hygroscopic nuclei will become well diluted and the effect becomes less pronounced. In addition, the radius of curvature also becomes large enough to approximate to a plane surface and, for large water droplets bigger

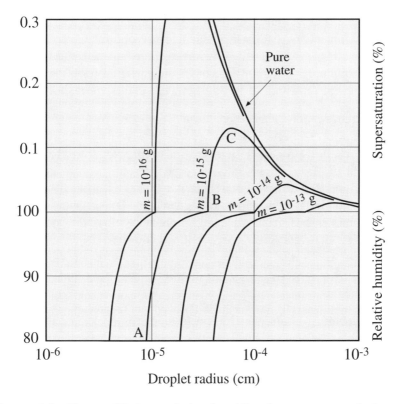

Figure 4.6 The equilibrium relative humidity (or supersaturation) as a function of the droplet radius for solution droplets containing the indicated mass of sodium chloride

than 1 μm, the difference between the two types of nuclei is no longer important. One interesting aspect of the difference in behaviour between hygroscopic and hydrophobic nuclei can be seen by comparing the growth of two small droplets. In an environment where the relative humidity is less than 100%, only hygroscopic nuclei will present possible condensation sites but, as the water droplet grows, the salt becomes more dilute and a higher relative humidity is required for further growth. This process is represented by path AB in Figure 4.6 and, once the droplet has reached the size dictated by the prevailing relative humidity, its growth will cease. Only if there is an increase in the relative humidity of the environmental air, or supersaturation is achieved, will further growth occur. An example of this process is the development of haze. A maritime haze layer will develop for a few hundred metres above the surface of the ocean, where salt nuclei are readily available because of bubble bursting at the sea surface. Haze over land may form as the result of the introduction of hygroscopic pollutants from industrial processes or by photochemical reactions of the gaseous car exhaust emissions.

In an environment where there are very few hygroscopic nuclei, supersaturation of the air, of the order of 1%, will be required to start the condensation process. However, once a droplet starts to grow, its radius will increase, the SVP (with respect to the curved surface) will decrease, and the droplet will find itself in a continually saturated environment and will carry on growing. Thus, once a critical supersaturation has been reached, represented by the point C in Figure 4.6, droplets will grow but, as the supersaturation curve approaches 100% relative humidity, the rate of growth will decrease. Water droplets larger than 10 μm in radius grow only very slowly by this process.

The availability of condensation nuclei depends on the rate of supply of suitable particles and on their rate of removal, either by gravitational settling or by precipitation known as 'washout'. Figure 4.7 shows the typical sources and sinks of atmospheric aerosols and the size distributions for the various types of aerosol. The small Aitken nuclei, less than 0.1 μm in radius, tend to be produced by smokes and gases, and are plentiful in industrial areas. They tend to be too small to be efficient as condensation nuclei on their own, but coagulation can produce larger nuclei, which are very efficient condensation nuclei. The majority of the aerosol is removed by 'washout', whilst the remainder is removed by dry deposition or gravitational settling. A particle 1 μm in diameter would take about one year to fall 1 km, whilst a 10 μm particle would take only 3.5 days, and thus gravitational settling is an important process only for the deposition of giant aerosols. For smaller Aitken nuclei, scavenging by precipitation will be the principal removal process.

The number of nuclei reduces from $10^{12}\,\text{m}^{-3}$ over industrial areas to about $10^8\,\text{m}^{-3}$ over the ocean, where the dominant nuclei are salt particles. These are principally produced by bursting bubbles in foam patches which are produced, in their turn, by breaking waves. The number of particles entering the marine atmosphere is dependent on the area of foam over the ocean, and this is determined by the sea state

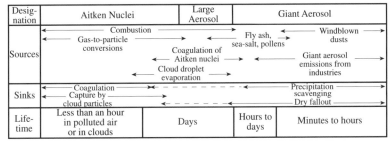

Figure 4.7 Schematic curves of aerosol surface area distributions for urban polluted air, continental air and marine air. Shown below the curves are the principal sources and sinks of atmospheric aerosols and estimates of their mean lifetimes in the troposphere

and the surface wind speed. The number of particles can vary from $10^6 \, \mathrm{m}^{-3}$ in low wind conditions to $10^9 \, \mathrm{m}^{-3}$ in storm-force conditions. Many of the larger particles will re-enter the ocean, but a small proportion will be taken aloft by atmospheric turbulence. These salt particles can have a very large range and they may be found in clouds many hundreds of kilometres away from the ocean. As a general rule, there are fewer condensation nuclei over the ocean than over the continents (Figure 4.7). The giant aerosols are produced mainly by 'mechanical' wind action at the ocean or land surface under storm conditions. Wind-blown sand, called aeolian, can be transported considerable distances, and red Saharan dust is occasionally reported in the British Isles and in the Caribbean, over 3000 km from the Sahara. In addition, large industrial plants produce considerable numbers of giant nuclei.

The typical size distribution of water droplets in clouds may now be considered. It may be obtained by exposing an oil-covered slide from an aircraft and measuring the droplet population with a microscope. A typical distribution is described by three parameters:

(i)   The mean radius.
(ii) The median radius.
(iii) The maximum radius.

Table 4.1 shows the statistics of drop distributions in various cloud types. In continental clouds there is a much larger number of droplets than in ocean clouds, because there is an increased population of large condensation nuclei over land. The amount of water vapour available for condensation is determined by large-scale processes, such as uplift, thus ensuring the liquid water contents of maritime and continental clouds are similar. Hence, the sizes of the droplets in continental clouds is smaller than in maritime clouds. It is noted from Table 4.1 that the maximum droplet size in maritime clouds can be twice as large as those in continental clouds. This observation is of critical

Table 4.1 Cloud drop distributions and liquid water content in typical maritime and continental clouds

Cloud type	Droplet number density (cm^{-3})	Range of radii of droplets (μm)	Most frequent radius of droplet (μm)	Liquid water content (g m^{-3})
Continental cumulus	420	3–10	6	0.4
Maritime cumulus	75	2–20	11	0.5
Cumulonimbus	72	2–100	5	2.5
Altostratus	450	1–13	4	0.6
Continental stratus	260	1–22	4	–
Maritime stratus	24	2–45	13	0.35

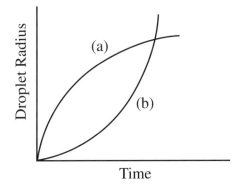

Figure 4.8   Schematic curves of droplet growth by (a) condensation from the vapour phase, and (b) coalescence of droplets

importance in the determination of the time scale for the growth of droplets in clouds.

Table 4.1 also shows that the maximum droplet size is related to the type of cloud. Weak vertical motions in the stratocumulus result in a reduction in the maximum size, whilst the strong vertical motions found in cumulonimbus result in large droplets. The water content of cumulonimbus clouds is much larger than in cumulus clouds because of the different surface-to-volume ratio of the two types of cloud. In cumulus cloud, usually 100 m to 1 km thick, this ratio is large and mixing of dry air across the surface of the cloud will significantly modify the water content. In the cumulonimbus, the ratio will be smaller because of the size of the cloud, typically 5–10 km thick, which implies that mixing across the surface of the cloud will have less effect in reducing the water content.

A qualitative discussion of the growth of a small water droplet in a supersaturated environment has already been given. This growth is dependent on (i) the supersaturation of the environment, and (ii) the size of the drop, and is related to the rate of diffusion of water vapour onto the droplet. As the droplet grows, the curvature effect on the SVP decreases, and therefore the rate of growth decreases. The growth of a small droplet, 1 μm in diameter, is depicted in Figure 4.8. It takes approximately 30 min for the droplet to reach 10 μm, a further $1\frac{1}{2}$ hours to reach 20 μm and 10 hours to grow to 40 μm. Even a 40 μm droplet is a long way short of the 200 μm of a respectable drizzle droplet. Clearly, the diffusion process cannot account for the growth of raindrops on the time scale of a shower cloud, which is typically $\frac{1}{2}$–2 hours. The additional process is the growth of raindrops by collision and coalescence. A droplet may collide with another drop, coalesce and increase its mass. Its vertical velocity will increase and it will sweep out more droplets at a faster rate, and it will soon reach a precipitation size. The process is non-linear in that initial growth will be slow because of the low terminal velocity of the small droplet, but the growth rate will become larger and larger as the droplet grows. This is shown in Figure 4.8b.

The rate of increase in mass of a droplet is given by:

$$\frac{\mathrm{d}M}{\mathrm{d}t} = \pi r_1{}^2 (v_1 - v_2) q_L E_c \qquad (4.15)$$

where r_1 is the radius of the droplet, $v_1 - v_2$ is the difference between the terminal velocities of the growing droplet and the surrounding cloud droplets, respectively, q_L is the liquid water content and E_c is the collision efficiency. Except in the initial stage of growth, $v_1 \gg v_2$, and substituting for $M = \frac{4}{3} \pi r_1^3 \rho_L$ into equation 4.15 yields:

$$\frac{\mathrm{d}r_1}{\mathrm{d}t} = \frac{v_1 q_L E_c}{4 \rho_L} \qquad (4.16)$$

The terminal velocity, v_1, for different droplets is shown in Table 4.2. A 10 μm water droplet has a terminal velocity of $10^{-2}\,\mathrm{m\,s}^{-1}$, whilst a drizzle droplet has a terminal velocity of about $\frac{1}{2}\,\mathrm{m\,s}^{-1}$ and a raindrop

Table 4.2 Terminal velocities of water droplets in still air at surface atmospheric pressure and temperature of 20°C

Radius of droplet (μm)	Terminal velocity in still air (cm s^{-1})
1	0.01
5	0.3
25	7.2
50	25.6
250	204
500	403
1000	649
2500	909

has a maximum terminal velocity of $9\,\mathrm{m\,s^{-1}}$. Thus, the rate of growth, which is proportional to the terminal velocity, will accelerate with droplet size. The growth rate is also dependent on the liquid water content of the cloud and the collision frequency. As already noted, cumulus and stratocumulus tend to have a lower water content than the larger cumulonimbus, and therefore the precipitation process will be more effective in the latter cloud. The collision frequency increases with droplet size and is low for droplets less than $20\,\mu\mathrm{m}$ in radius. When the cloud droplets are small, they tend to follow streamlines around the collector droplets and do not get captured. For larger droplets of $70\,\mu\mathrm{m}$ radius or more, turbulent wakes develop behind the droplet, drawing in more droplets than were originally in the volume swept out.

The production of precipitation from a water-droplet cloud is a non-linear process. In a given distribution of cloud droplets, only a few droplets will be bigger than $20\,\mu\mathrm{m}$, and thus be sufficient in size to grow efficiently by collision and coalescence. At first, the process will be slow but, because of the non-linearity of the process (Figure 4.8), the droplet will soon grow quickly to the size of a raindrop. In a maritime cumulus, with droplet concentrations of $100\,\mathrm{cm^{-3}}$, droplets will grow from 25 to $1000\,\mu\mathrm{m}$ within 10 min. Of course, only a small proportion at the extreme end of the droplet population will be involved in the precipitation process. The differences between the size distributions in maritime and continental clouds has already been noted, as well as the fact that in many cases r_{max} is less in continental than in maritime clouds. A continental cloud will therefore take longer to produce rain than the corresponding maritime cloud, and it has been calculated that a maritime cumulus cloud 1.5 km deep will produce rain within 30 min of its formation, whilst similar continental cloud will take more than an hour. This illustrates how dramatic an influence the concentration of condensation nuclei has on the rate of growth of water droplets, and hence on the formation of precipitation.

The growth of a shower cloud is dependent on:

(i)   The cloud droplet size distribution.
(ii) The liquid water content of the cloud.
(iii) The thickness of the cloud.
(iv) The updraft within the cloud, which is essential for the growth of cloud droplets into raindrops.

Small droplets have small terminal velocities and will be maintained in the cloudy environment for longer than larger droplets, provided that they are not ejected through the top of the cloud. As the updraft increases, the size of the final raindrop increases up to a maximum size of 5 mm and, at this stage, the raindrop will break up into smaller droplets. A maritime cumulus will have to be at least 1 km thick to produce light precipitation. However, drizzle may be produced by stratus cloud whose thickness is less than 1 km because the weak vertical motion within such clouds allows the droplets to be maintained in the cloudy air for a sufficient time to reach drizzle drop size.

So far, the growth of precipitation in clouds where the temperatures are above 0°C has been discussed. These are known as warm clouds, and are found as tropical cumulus clouds and as low-level clouds in middle latitudes. However, below 0°C, additional physical processes come into play as a result of the introduction of the ice phase. Consider a cloud of liquid water droplets with a mean size of 10 μm. As the droplets are cooled below 0°C, only a few droplets will freeze and the remainder will be in a supercooled liquid state. As the temperature is reduced still further, the number of ice crystals will increase, but only when the temperature reaches −39°C will all the water droplets have frozen. At this temperature, spontaneous nucleation will occur, caused by the chance aggregation of water molecules into an ice embryo. This process is analogous to the condensation of water vapour in a cloud chamber in the absence of ions or other particles. However, the observation that ice crystals are formed at temperatures higher than −39°C indicates the existence of ice nuclei. These ice nuclei are different from condensation nuclei in two important respects. First, the most successful ice nuclei tend to have hexagonal molecular structures similar to ice. The most common types of ice nuclei are clay minerals, obtained from the soil, and organic materials, particularly decayed plant matter and derivatives of phytoplankton from the ocean. Secondly, the ice nuclei have a threshold temperature at which they become effective and below which they act as preferential sites for ice nucleation. For clay particles, this temperature is −15°C, whilst organic nuclei may have threshold temperatures as high as −4°C. Thus, the presence of ice nuclei only has a significant effect below −4°C, and this explains the observation that clouds between 0 and −4°C exist in the supercooled liquid state.

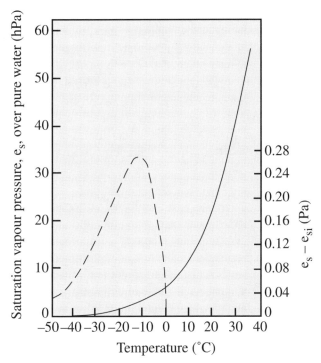

Figure 4.9 Variations of the saturation vapour pressure (e_s) with temperature over a plane surface of pure water (— —), and the difference between (e_s) and the saturation vapour pressure over a plane surface of ice e_{si} (——)

The processes which lead to the growth of ice crystals in a supercooled cloud will now be considered. Figure 4.9 shows the SVP over a plane water and a plane ice surface. It is noted that the SVP over ice is always lower than over liquid water, except at 0°C, and that water vapour in equilibrium with a water surface will therefore always be supersaturated with respect to an ice surface below 0°C. The difference in SVP with respect to ice and water reaches a maximum at −12°C. Here, in a saturated environment with respect to supercooled liquid water, the vapour would have a supersaturation of 10% with respect to ice. In a mixed cloud, consisting of supercooled liquid water droplets and ice crystals, the ice crystals will grow at the expense of the water droplets. This process is known as the Bergeron process, after the Norwegian cloud physicist who first noted its existence. It is most effective in clouds where the temperature is between −10 and −30°C (i.e. where the difference in SVP is greatest). Between 0 and −10°C there are two few ice nuclei present for the mechanism of ice crystal formation to be effective, whilst below -30°C there are too many ice crystals competing for too few liquid water droplets for the process to be significant. The growth rates of the ice crystals in mixed clouds are much greater than the growth rates of water droplets in warm clouds because of the higher supersaturations occurring in mixed clouds compared with those in warm clouds, typically 10% supersaturation compared with less than 1% supersaturation. For example, an ice crystal can grow to 1000 μm in diameter within 30 min by the Bergeron process.

An ice crystal can grow by two additional processes other than the Bergeron process. First, it can grow by aggregation of ice crystals, because supercooled water films attached to an ice crystal can act as binding for other ice crystals as a result of surface tension between the water and the ice. The thin film of supercooled water then freezes and the crystal grows. The process is most effective between 0 and -10°C, where there are a large number of supercooled water droplets and freezing does not occur immediately on impact. A snowflake may grow from 1 mm to 1 cm in about 30 min by aggregation, and the largest snowflakes occur at temperatures close to 0°C. At lower temperatures, supercooled water droplets may freeze on impact with ice crystals to produce a rimed snowflake. This process is known as accretion, and it is responsible for the growth of ice crystals between -10 and -30°C. It is most effective in clouds such as cumulonimbus where the liquid water content is high. In contrast, the Bergeron process is most effective in small cumulus and stratocumulus clouds where the water content is low. Accretion and the riming of ice particles can produce soft hail and ice pellets and, more exceptionally, hailstones. The accretion process has many similarities with the collision–coalescence mechanism in that it is non-linear due to the increase in terminal velocity with size. For crystals formed by aggregation rather than accretion, the terminal velocity does not tend to increase with the size of the crystal because of the crystal's large surface-to-mass ratio. The drag force on a crystal increases in proportion to its mass, and hence an unrimed crystal 2 mm in diameter may have a terminal velocity of 0.5 m s^{-1}, whilst a rimed ice pellet of similar diameter would have a terminal velocity of 1.8 m s^{-1}. Therefore, an ice pellet falling through a cloud of high water content could grow very quickly by accretion. Riming, or accretion, can also cause a rapid build up of ice on solid surfaces, such as aircraft wings and television transmitters. In such cases, impact freezing will generally occur at all temperatures below 0ºC, and it is especially important between 0 and −4ºC where large numbers of supercooled water droplets exist. In March 1969 a television transmitter in Yorkshire, England, collapsed as a result of excessive loading of rimed ice on its structure.

Figure 4.10 shows a simplified diagram of the precipitation processes occurring in layered clouds associated with a typical middle-latitude (extra-tropical) cyclone. The cirrus and cirrostratus clouds are composed entirely of ice crystals, whose growth is dependent on the number of ice nuclei and the amount of water vapour available. The freezing process in cirrus clouds produces ice splinters, which thereby increase the number of ice nuclei. These ice splinters have fall speeds of approximately $0.5 \, \text{m s}^{-1}$ and they can, in turn, seed the supercooled clouds below. Ice crystals do not evaporate as readily as water droplets and they may fall through layers of clear air up to 500 m thick before finally evaporating. This can occasionally be observed in cirrus, where fall streaks give a characteristic hook shape to the cloud. In the middle-level clouds, such as altostratus, the clouds are mixed, and both Bergeron and accretion processes are important. At distances of 1.5 km or less above the freezing level, aggregation becomes the dominant process and, depending on the temperature of the lower layer of the atmosphere, precipitation will fall as either rain or snow.

The growth of hailstones is an extreme example of the accretion process and it occurs in deep convective clouds, which have both a high water content and a very strong updraught. Under exceptional circumstances, hailstones as large as 13 cm in diameter (about the same size as a grapefruit) have been recorded, although most hailstones are less than 1 cm across. Even in the United Kingdom, hailstones as large as 7 cm have been recorded in summer storms. Hailstones are formed in only a small region of a cumulonimbus cloud on the forward side of the updraught region, as shown in Figure 4.11. The hailstones usually occur for a few minutes in the course of a storm which may last for up to an hour. When a thin-section of a hailstone is viewed in transmitted light, a banded appearance is usually seen, with alternating layers of clear and opaque ice. This banded structure is the result of the impact of supercooled water droplets at different heights within the cloud. In the upper levels of the cloud, where the temperature range is from −20 to −40ºC, water droplets are not numerous, freezing occurs on impact

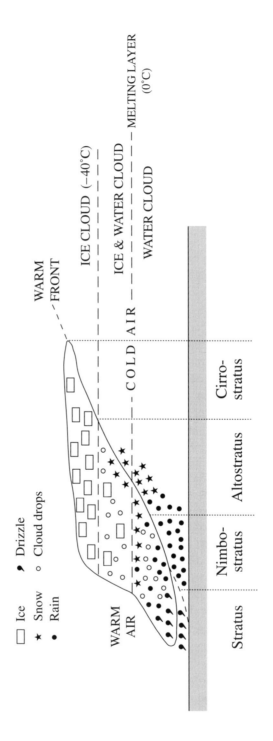

Figure 4.10 The Norwegian model of the microphysical structure of a warm front. The horizontal scale is typically 100 to 200 km, the vertical scale is about 10 km and the uplift of the warm air over the cold air is about 0.1 m s^{-1}

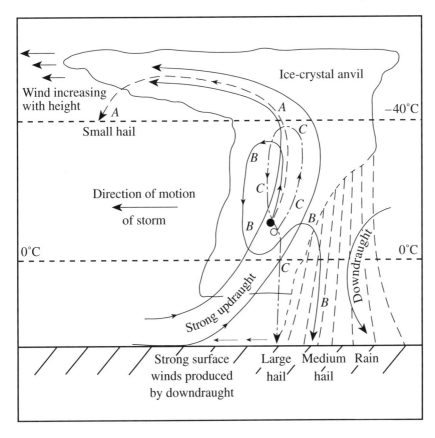

Figure 4.11   The structure of a hailstorm. The horizontal and vertical scales are typically 10 km, though the anvil may spread many times this distance. Large hail occurs in the region of cloud where recirculation in the updraught is possible

and air is trapped between the layers, thus giving a milky appearance. At lower levels in the cloud, where the temperature range is from -20 to 0°C, there are more water droplets, and the hailstone collects water droplets at too fast a rate to be able to dissipate the latent heat of fusion by conduction to the air, as required if spontaneous freezing is to occur. The surface of the hailstone becomes wet and it eventually freezes to form a layer of clear ice. A hailstone will often require a number of cycles within the cloud to grow to a reasonable size and, for this to occur, the updraught must exceed $10 \, \mathrm{m \, s^{-1}}$. As a general rule, the maximum updraught in the upper troposphere is equivalent to the final terminal velocity of the hailstone. A hailstone 1 cm in diameter has a terminal fall speed of $25 \, \text{m s}^{-1}$, whilst a 10 cm hailstone has a fall speed of $65 \, \text{m s}^{-1}$. The latter size of hailstone would have a mass of about 0.5 kg and would be able to shatter glass, break roof tiles, and indent cars and caravans.

4.6   Macroscopic processes in cloud formation

One advantage that the meteorologist has over the oceanographer is the ability to see the formation of clouds, and to use his or her direct observations to make deductions about atmospheric motions and processes. From a microphysical point of view, it is possible to distinguish three types of cloud. At temperatures below $-39°C$ all clouds are composed of ice crystals. These clouds are known as 'cirrus' and they tend to have fuzzy edges because of the slow evaporation of ice crystals ejected from the cloud, as discussed in Section 4.5. These cirrus clouds are optically thin and they allow solar radiation to penetrate to lower levels in the atmosphere. The alignment of these clouds gives information on the direction of the wind and the fall streaks allow an estimate of the vertical wind variation to be made. The hexagonal structure of the ice crystals gives rise to preferred directions of scattering, and thus to a variety of optical phenomena.

At temperatures between -39 and $0°C$, supercooled water droplets and ice crystals co-exist and, in this temperature range, the clouds are therefore termed 'mixed clouds'. At temperatures colder than $-30°C$, the water droplets are outnumbered by the ice crystals and so mixed clouds tend to behave optically like pure ice clouds. Between -10 and $0°C$, ice crystals are few in number, and therefore these clouds behave optically as if they were water clouds. Typical mixed clouds are altocumulus and altostratus, which form at mid-levels in the troposphere. In thin mixed clouds with small water droplets, coloured diffraction rings can appear round the Sun or Moon, forming a corona.

At temperatures above $0°C$, clouds are composed solely of water droplets. Such clouds include small cumulus, stratus and stratocumulus, and they exist at low levels in the troposphere. Water-droplet clouds tend to have distinct edges due to the rapid evaporation of ejected water droplets in an unsaturated environment. Such sharp delineation is often seen in growing cumulus.

However, the microscopic classification of clouds has its limitations in ascertaining cloud processes since, for example, it does not distinguish between cumulus- and stratus-type clouds. These two types of cloud, which are the basis of the descriptive classification of clouds formulated in the nineteenth century by Luke Howard, give information about the stability of the atmosphere. Layered, or stratus, clouds form in a locally stable environment, whilst lumpy, or cumulus, clouds are indicative of vertical instabilities. Low-level stratus-type clouds can be formed by:

(i)   Advection of warm, moist air over a cooler surface. This may cause advection fog, particularly over cold currents and snow-covered surfaces.
(ii) Radiation cooling of a moist air mass during its transition from low to high latitudes.
(iii) Gentle uplift of a stable air mass, either by a topographic feature or in a large frontal region associated with an extra-tropical cyclone.

On the other hand, cumulus-type clouds are formed in a locally unstable environment produced by advection of cool, drier air over a warm (moist) surface. In summer, a stable air mass may be rendered unstable, in the lower 1 km, by surface heating over land. This gives rise to small cumulus, known as fair-weather cumulus, providing that the convection reaches the condensation level. Over the ocean, polar air masses moving equatorwards over warmer surface waters can give rise to convection. The depth of the convection depends upon the dynamical motions maintaining the stability of the troposphere. For example, in an atmosphere of weak vertical stability, such as occurs in the rear of an extra-tropical cyclone or in the ITCZ, convection may extend from the sea surface to the top of the troposphere. In such cases, deep convective cumulonimbus clouds will form. These clouds will consist of water droplets in the lower troposphere and predominantly of ice crystals in the cirrus anvil at the top of the troposphere. In the trade-wind region, or in a middle-latitude anticyclone where the air has strong vertical stability, the convection will be limited and cumulus will form in the moist boundary layer. However, progressive warming of a trade-wind air mass as it crosses the ocean may be sufficient to produce convection 2–3 km deep, which is sufficient to produce showers. A less commonly observed convective cloud is termed 'arctic smoke' and it is formed by the movement of a very cold air mass over a warmer sea surface. The layer of air adjacent to the sea is strongly unstable and transfers heat and, more importantly, water vapour into the air mass, and this water vapour condenses on mixing with its cold environment. This process quickly warms the arctic air mass.

Instability may be induced not only by surface heating, but also by the advection of colder air over a warm air mass. This type of instability often occurs in summer when a cooler maritime air mass is advected over a very warm surface air mass of continental origin. A notable feature of this type of instability is the appearance of lines of convective cloud in the middle troposphere, which are known as castellanus clouds because of their 'castle-like' appearance.

Unstable and stable air masses may co-exist and produce a variety of clouds. As noted earlier, a stable air mass may be rendered unstable by surface heating, which in turn leads to the development of cumulus. If there is sufficient moisture present and if the surface heating is strong enough, the cumulus may develop to a depth of 1 km, which may be enough for the development of an isolated light shower. However, the stability of the air above the cloud may be strong enough to limit further upward development of the cloud which will, therefore, spread laterally. This cloud, known as stratocumulus, will often produce a complete layer of cloud cover, which will persist until the surface heating decreases or until sufficient dry air from above the inversion is mixed into the clouds to promote evaporation. This type of cloud often has a diurnal cycle, with maximum thickness occurring in the early afternoon, and usually clearing at dusk when the surface heating is reduced. Alternatively, a layered cloud may break up into convective-type cloud because of solar radiational heating. As discussed in Chapter 2, solar radiation penetrates a cloud to a considerably deeper depth than long-wave planetary radiation. Thus, the top of a cloud will cool by long-wave radiative emission into space, whilst the bulk of the cloud mass will heat up by solar radiation absorption. This process results in an increase in the vertical temperature gradient within the cloud, which in turn leads to instability. This phenomenon is more often observed in thin clouds like altostratus or cirrostratus, which are transformed into altocumulus and cirrocumulus.

The mixing of air masses is another important macroscopic cloud process. The importance of the mixing of cloudy air with dry air to dissipate cloud has already been mentioned, but if two air masses close

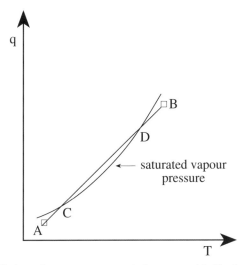

Figure 4.12 Mixing of two unsaturated air masses (A, B). A mixture lying on the line CD will become saturated

to saturation mix, clouds may form. The reason for this may be understood by referring to Figure 4.12. If air mass A and air mass B are mixed, then the resultant air mass will lie along the line AB and its exact location will depend on the relative properties of the two air masses. The saturation mixing ratio does not vary linearly with temperature and so the points lying along the line CD will represent saturated or supersaturated air, and thus condensation would be expected. Only for mixtures lying on the lines AC and BD will the air remain unsaturated. A common example of this phenomenon is the emission of hot, moist exhaust gases from a car or an aeroplane into a cold, moist environment. In this case, because of the low environmental temperature, only a small amount of water vapour is required for condensation.

Vertical mixing of air masses may also result in cloud formation. One particular example is the formation of stratocumulus at the base of the inversion in an anticyclonic system. In this case, mixing of cool moist surface air with warm dry air aloft will result in a decrease in temperature, but an increase of mixing ratio at the top of the layer. The converse will occur in the lower layer, thus resulting in condensation in the upper layer. This process is often responsible for the development of stratocumulus over the sea, when the sea temperature is lower than the temperature of the air above it. In a winter anticyclone, over a moist land surface, a thick layer of stratocumulus may persist for many days, giving rise to conditions known as 'anticyclonic gloom'. Clearly, the formation of this low-level cloud is dependent on a number of processes. The upward flux of water vapour from the surface and vertical mixing will tend to enhance cloud formation, whilst the downward mixing of warmer, dry air from above the inversion will tend to dry out the boundary layer and dissipate the cloud.

Vertical mixing is also important in the development of surface fogs. These may develop by long-wave radiational cooling of the land at night which, if sufficient, will lower the ground temperature to the dew point of the overlying air. In light winds, only the surface layers of air will reach the dew point and a shallow layer of fog will form. However, with stronger vertical mixing, the depth of air cooled to the dew point will increase, and hence a thicker layer of fog will develop. Another example is the formation of advection fog, formed by the flow of warm, moist air over a cold surface, such as a cold ocean current or a snow-covered land surface. In this case, vertical mixing will tend to enhance the depth of the fog.

The third, and most important, contribution to the formation of clouds is adiabatic cooling. A spectacular example is the formation of funnel cloud in a tornado vortex. The low pressure of such a vortex is often sufficient to adiabatically cool the air to the condensation point. For example, a fall of 5 kPa in the pressure will cause an adiabatic cooling of 4°C at the surface. But by far the most common cause of condensation is adiabatic cooling by vertical uplift. This vertical motion may be caused by the forced lifting of an air mass over a topographical feature, by convection (as discussed previously) or by large-scale ascent associated with frontal or synoptic scale features. The lifting of air masses over topography may give rise to relatively strong vertical motions of $1\,\mathrm{m\,s^{-1}}$ and to the development of 'local' clouds over the topographic feature. In a stable environment, wave clouds may form over the mountain and downstream of the mountain for up to 100 km. Clouds formed by large ascents in fronts tend to be associated with weaker vertical motions of approximately $0.1\,\mathrm{m\,s^{-1}}$, and therefore the clouds tend to be layered. However, regions of stronger uplift, associated with rain bands in a frontal system, may produce more convective-type cloud.

5

Global Budgets of Heat, Water and Salt

5.1 The measurement of heat budgets at the surface

The net heat flux, Q_T, across a horizontal surface is given by

$$Q_T = Q_S(1 - \alpha) - Q_N - Q_E - Q_H \tag{5.1}$$

where Q_S is the downward solar radiation flux, α is the surface albedo, Q_N is the net upward long-wave or planetary radiation flux, Q_E is the heat lost by the surface due to the latent heat of evaporation and Q_H is the upward turbulent heat flux. Thus, four separate quantities must be determined in order to calculate the total heat flux across an ocean or land surface, and the method used over land and sea to make the measurements over limited areas will now be discussed. The net radiation flux, Q_R, is given by $Q_R = Q_S(1 - \alpha) - Q_N$.

The solar radiation flux, Q_S, can be determined using a solarimeter, which measures both the direct and diffuse radiation between 0.35 and 2.8 μm, i.e. over most of the solar spectrum. The solarimeter consists of two elements, one blackened to both solar and long-wave radiation, and the other covered with a white paint which has a high short-wave reflectivity. The temperature difference between the two elements, measured by a series of thermocouples, is proportional to the incident short-wave radiation. A glass dome covering the sensor elements gives the instrument protection from the weather, and it can be mounted on a ship and give reliable readings for long periods of time. The accuracy of the instrument is about 2%.

The net long-wave flux, Q_N, can be determined from measurements of the total net radiation and the solar radiation. A net radiometer consists of two blackened thermopiles, one pointing upwards and the other pointing downwards. The voltage difference between the two thermopiles is a measure of the net radiation across the surface, which may thus be determined with an accuracy of about 2.5%. The thermopiles are enclosed in a polythene dome which transmits both long- and short-wave radiation. The dome is ventilated with dry nitrogen gas to avoid water-vapour absorption and condensation. Measurements of net long-wave radiation are obtained routinely at many meteorological stations, but measurements at sea are more difficult. It is necessary to avoid contamination by reflections from the ship, and thus a boom is required to obtain uncontaminated fluxes. However, providing that the sea surface temperature is known, the upward flux of long-wave radiation can be calculated from Stefan's law, $E = \epsilon \sigma T^4$, with an emissivity, ϵ, of 0.97, and thus only the downward long-wave flux from the sky needs to be measured. This can be done using an upward-looking radiometer.

One of the most accurate determinations of the latent heat flux, Q_E, and the upward turbulent heat flux, Q_H, is by the eddy correlation method. Over both land and sea, the upward flux of heat (and water vapour) is caused by turbulent motions close to the surface. Figure 5.1 shows the fluctuations of vertical velocity and temperature for a 2 min record at 2 m over land. Despite the randomness of the fluctuations, it can be seen that high temperatures are generally associated with the rising motion, whilst low temperatures are associated with the descending motion. In this situation, a mean vertical gradient of temperature exists with high temperature close to the surface. Fluctuations in vertical velocity, associated with convection or surface friction, cause an exchange of parcels of air across the gradient and this, in turn, gives rise to an upward flux of heat. The correlation coefficient between the vertical velocity and temperature gives an accurate measurement of the heat flux at the level of instrumentation. Providing that there is no loss of heat between the surface and the instrument, the correlation coefficient will give a measurement of the surface turbulent heat flux. In a similar manner, the correlation coefficient between the vertical velocity and humidity fluctuations allows a determination of the latent heat flux. The turbulent heat flux is given by:

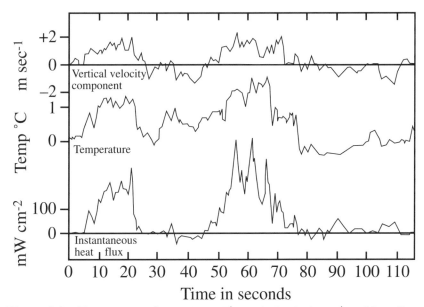

Figure 5.1 Temperature fluctuations, T', vertical velocity, w', and heat flux, $w'T'$, at 2 m above the Earth's surface

$$Q_H = \rho C_p \overline{w\theta} \qquad (5.2)$$

and the latent heat flux by:

$$Q_E = L\rho \overline{wq} \qquad (5.3)$$

where the overbar denotes a time average of the instantaneous values of the vertical velocity, w, the mixing ratio, q, and the potential temperature, θ, over a period of 10–60 min.

This method requires sophisticated and accurate instrumentation in order to measure the high-frequency variations in w, q and θ. In one 10 min average, about 10 000 observations are required to obtain reliable fluxes. Over the sea, there are more difficulties because of the ship's movements and the contamination of the instruments by salt spray. For these reasons, the above technique is only practicable for the determination of latent heat and turbulent heat fluxes from a research ship, for limited periods and in light to moderate wind conditions. Another direct method, which can be applied from ships in strong wind conditions, is the dissipation method. This uses the high-frequency wind observations to determine the dissipation of turbulent kinetic energy. This measurement can be used directly to obtain the surface wind stress or it may be combined with additional high-frequency temperature and humidity measurements to determine turbulent heat and latent heat fluxes. These methods are very useful for the calibration of other, simpler formulae which are more widely used.

More common methods of measuring the latent heat and turbulent heat fluxes rely on the measurement of the mean vertical gradients of humidity and temperature which can be obtained, for example, using a standard ventilated psychrometer. By analogy with heat conduction in solids, the flux of latent heat can be written as:

$$Q_E = -\rho L K_E \frac{\Delta q}{\Delta z} \qquad (5.4)$$

and the flux of turbulent heat as:

$$Q_H = -\rho C_p K_H \frac{\Delta \theta}{\Delta z} \qquad (5.5)$$

where $\Delta q/\Delta z$ and $\Delta \theta/\Delta z$ are the vertical gradients of the water-vapour mixing ratio and the potential temperature, respectively. The coefficients K_E and K_H are known as the eddy exchange coefficients and have to be determined for each set of observations, unlike the thermal conduction coefficients which remain constant over reasonably wide ranges of temperature and pressure. In order to calculate the exchange coefficients, an equation for the tangential stress, τ, must be introduced:

$$\tau = \rho K_M \frac{\Delta u}{\Delta z} \qquad (5.6)$$

where K_M is the eddy exchange coefficient for momentum and $\Delta u/\Delta z$ is the vertical gradient of the wind speed. Although the eddy coefficients all vary with height above the surface and with the stability of the atmosphere, it is reasonable to assume that, at a given height and time, the mixing processes for humidity, temperature and momentum will be similar, and vertical eddy circulations will carry parcels of momentum, temperature and humidity simultaneously. In this case, $K_E = K_H = K_M = K$, and substituting for K from equation 5.6 into 5.4 and 5.5:

$$Q_E = L_T \frac{\Delta q}{\Delta u} \tag{5.7}$$

and

$$Q_H = C_p T \frac{\Delta \theta}{\Delta u} \tag{5.8}$$

From work on aerodynamic boundary layers, the surface stress, T_0, is known to be given by:

$$T_0 = \rho C_D U 10^2 \tag{5.9}$$

where C_D is the drag coefficient and U_{10} is the wind speed 10 m above the surface.

For a surface stress, the wind speed at the surface is zero and so $\Delta u = U_{10}$. In addition:

$$\Delta q = q_A - q(\theta_S) \tag{5.10}$$

and

$$\Delta \theta = \theta_A - \theta_S \tag{5.11}$$

where θ_A and q_A are the potential air temperature and mixing ratio, respectively, 10 m above the surface, and θ_S is the sea surface temperature. $q(\theta_S)$ is the surface mixing ratio.

Substitution of equation 5.9 into 5.7 and 5.8, and using 5.10 and 5.11 yields:

$$Q_E = -\rho L C_D U_{10}(q_A - q(\theta_S)) \tag{5.12}$$

$$Q_H = -\rho C_p C_D U_{10}(\theta_A - \theta_S) \tag{5.13}$$

Over the sea surface, $q(\theta_S) = q_s(\theta_S)$, where q_s is the saturation mixing ratio. These relationships are most applicable to the computation of fluxes over the sea surface. They require only observations of U_{10}, q_A, θ_A and θ_S, and these can be obtained from ships' reports as a matter of routine. However, they have the disadvantage that the drag coefficient, C_D, is not a constant, being dependent on both vertical stability and on wind speed. It is fortunate that, over large areas of the ocean, the atmospheric boundary layer near to the sea surface is often close to neutral stability and that the dependence on wind speed only becomes noticeable at high wind speeds. Comparisons between the eddy method and the aerodynamic equations have shown that the latter are accurate to about 10% in moderate wind conditions over the ocean.

Over the land, the aerodynamic equations 5.12 and 5.13 are not as useful because the drag coefficient will vary with the type of surface. In this case, in order to use equations 5.7 and 5.8, τ has to be determined from the wind profile from the surface to about 10 m, which, in neutral stability conditions, is given by:

$$U = \frac{U_*}{k} \ln \left(\frac{z}{z_0} \right) \tag{5.14}$$

where U_* is the friction velocity and k is the von Karman constant which has a value of 0.4. z_0 is the roughness coefficient. The friction velocity is defined by:

$$\tau = \rho U_*{}^2 \tag{5.15}$$

Thus, from a vertical profile of U, the gradient will determine the friction velocity, U_*, and thence the stress, τ. The roughness coefficient, z_0, can also be determined from the velocity profile. Typical values of z_0 vary from 0.01 cm over smooth sand to about 100 cm over scrub country.

Apart from the use of the aerodynamic method, all of the previous methods rely on instrumentation which is only available on research ships, weather ships and some automatic ocean buoys. If it is assumed that there are, at most, 100 specially equipped research vessels at sea at any one time (and this is probably a considerable overestimate), then, provided that they were evenly distributed, one observation would be made for each 2000 km square of ocean. This is an unsatisfactory data set for the consideration of surface fluxes on a global scale. The global maps of surface fluxes in the scientific literature are compiled from the merchant vessel observations rather than those from research ships. As noted, the bulk aerodynamical method is suitable for estimating the turbulent heat flux and the latent heat flux from standard weather reports, but even here there are problems. The height of the wind observations will vary from ship to ship, and wet-bulb thermometers can be contaminated by salt spray and may, therefore, give a systematic error in humidity if they are not regularly maintained. In addition, the water temperature is measured at the engine intake point, which may be a few metres below the sea surface, and the recorded temperature may be systematically different from the sea surface temperature, particularly in low wind conditions when vertical mixing is reduced.

The problems posed by the global measurement of solar radiation and long-wave radiation are rather more difficult to overcome than those mentioned above because of the variable effect of cloud. In clear sky conditions, provided that the aerosol is low, both the solar radiation and the long-wave radiation can be determined from empirical formulae. The solar radiation impinging on the top of the atmosphere can be calculated for a given latitude and time of year, as demonstrated in Chapter 3. Allowance can be made for the depletion of the solar radiation by atmospheric gases and a 'standard' maritime aerosol to obtain a value for the solar radiation incident at the surface. However, recent research has shown that horizontal variations of aerosol can cause variations of as much as 5% in the solar radiation. The simplest formulae which include the effect of cloud are of the form:

$$Q_S = Q_C(1 - \alpha_c C) \qquad (5.16)$$

where Q_C is the daily averaged clear-sky solar radiation, C is the cloud fraction and α_c is typically 0.7. The use of this formula, whilst reasonable for the determination of monthly averages, can give errors of 50% in estimating daily averages. Substantial improvements on equation 5.16 can be obtained by using information on the type of cloud and the height of the cloud base, which are reported regularly in standard meteorological observations by ships.

Lumb (1964) showed that a more accurate formula for estimating hourly values of solar radiation is given by:

$$Q = Q^*(a_1s + b_1s^2) \qquad (5.17)$$

where s is the sine of the solar altitude and Q^* is the solar constant. a_1 and b_1 are coefficients which depend on nine categories of cloud conditions, which involve type, amount and cloud-base height. This formula is accurate to $\pm10\%$ for estimating the daily averaged solar radiation between 45 and 65°N. Recent work in the north-eastern Atlantic Ocean has shown that 12-day average solar radiation estimates from equation 5.17 agree with direct measurements to within 5%.

Net long-wave radiation under clear skies can be estimated from a formula proposed by Brunt (1932):

$$Q_{NC} = \epsilon \sigma T_S{}^4 (0.35 - 0.05\sqrt{e_a}) \qquad (5.18)$$

where ϵ is the emissivity of the sea surface ($\epsilon = 0.97$), T_S is the surface temperature measured in kelvin, σ is the Stefan–Boltzmann constant and e_a is the vapour pressure of the atmosphere at 10 m. The dependence on humidity is due to the re-emission of the long-wave radiation back to the surface by atmospheric water vapour. The majority of the re-emission comes from the first 100 m above the surface and the humidity at 10 m can, therefore, be used to estimate the downward long-wave radiation. There are two points to note about formula 5.18. First, over the observed range of sea surface temperatures, the net long-wave flux, assuming saturated air and clear skies, only varies from $70\,W\,m^{-2}$ at $271\,K$ to $23\,W\,m^{-2}$ at $300\,K$, which is much smaller than the latitudinal variation of solar radiation. Secondly, the net long-wave radiation decreases with increasing temperature because the downward long-wave radiation, which is proportional to the vapour pressure, increases more quickly with surface temperature than the upward radiation emitted by the surface. As with solar radiation, a cloud factor may be introduced in the form:

$$Q_N = Q_{NC}(1 - \alpha_c C) \qquad (5.19)$$

where C is the cloud fraction and $\alpha_c = 0.8$. Some formulae allow a variable α_c, dependent on the cloud type. For example, for low-level cloud like stratocumulus, $\alpha_c = 0.9$, whilst for high-level cloud such as cirrus, $\alpha_c = 0.25$. It is noted that, with total cover of low-level cloud, the net long-wave flux is approximately 10% of the net flux in clear-sky conditions, and so the cloud cover will further reduce the latitudinal variation of the long-wave radiation budget. Net radiation can be estimated from equation 5.18 with an accuracy of about 10%.

5.2   Observations of surface heat budgets

Table 5.1 shows the annually averaged components of the heat budget for land and ocean. It is noted that, over the annual cycle, the

Table 5.1   Annual surface heat budgets ($W m^{-2}$) over ocean and land. Q_R is the net radiation budget, Q_E is the latent heat of evaporation, Q_H is the upward conduction of heat and Q_T is the net heat flux into the ocean

	Ocean				Land		
Latitude zone	Q_R	Q_E	Q_H	Q_T	Q_R	Q_E	Q_H
50–70°N	35	48	21	−34	33	22	11
30–50°N	90	93	18	−21	70	31	39
10–30°N	155	137	10	8	94	33	61
10°S–10°N	154	110	5	39	96	66	30
30–10°S	143	137	8	−2	96	46	50
50–30°S	93	90	11	−8	69	33	36

net heat flux into the land will be close to zero because the heat gained in the summer period will be lost to the atmosphere during the winter. Over the ocean, there is a net gain of heat between 20°S and 20°N, and at higher latitudes there is a net heat loss. These heat deficits and surpluses must be compensated for by a transfer of heat from the equatorial zones to the middle and high latitudes via ocean currents.

The annual radiation budgets at lower latitudes generally show more heat going into the ocean than into the land. This observation is related to two factors. First, the albedo of the sea surface is much lower on average than that of the land at corresponding latitudes, especially the sub-tropical deserts, and therefore more solar radiation is absorbed by the ocean. Secondly, the surface temperatures of the tropical ocean are lower than those observed over land, and therefore the net long-wave radiation losses are less. Over the ocean, the largest net loss of heat is by evaporation; the turbulent heat conduction accounting for only 10% on average of the net heat loss. In the equatorial zone, Q_H contributes only 5% to the total heat loss, whilst at higher latitudes it contributes about 30%. Over land, both latent heat and heat conduction contribute equally to the overall net heat loss. In the sub-tropical deserts, heat conduction dominates over evaporation, whilst the opposite is true in the equatorial zone and over the wetter middle latitudes.

Figure 5.2 (a) Annual average heat gain, Q_T (W m^{-2}), for the North Atlantic

135

Figure 5.2 (b) Annual average turbulent heat flux, Q_H (W m^{-2})

136

Figure 5.2 (c) Annual average evaporation, E (cm yr^{-1}). Evaporation rate should be multiplied by 1.26 to get approximate heat flux, Q_E, in W m^{-2}

Figure 5.2 (d) Annual net radiation exchange, Q_R (W m^{-2})

To gain an impression of the longitudinal variations in surface heat budgets over the ocean, the annual average components are shown in Figure 5.2 for the North Atlantic, calculated from 25 years' observations by merchant vessels. Large evaporative heat losses of over $200\,W\,m^{-2}$ are seen over the Gulf Stream and these losses are twice as large as the evaporative losses in the sub-tropics. Furthermore, large turbulent heat losses are seen in the western Atlantic. In contrast, the radiation budget (Q_R) is rather zonal in character. Net heat gain occurs in the southern and eastern Atlantic, whilst net heat loss occurs in the northern and western Atlantic. The large heat losses in the western Atlantic are caused by a combination of the high sea surface temperatures and the frequent outbreaks of cold, dry air from the North American continent in winter. In the eastern Atlantic, the sea surface temperature is lower, especially in the cold upwelling zones off north-west Africa, and therefore evaporation and turbulent heat flux are reduced. The cold upwelling zone along the equator can easily be identified by the suppression of evaporation and the large net heat flux into the ocean.

It will now be shown how the distribution of the net heat flux across the ocean surface can be related to the transport of heat in the ocean. First, consider the heat budget for a slab of ocean extending across an ocean basin and over the whole depth. For the conservation of heat:

$$S_O + T_{i+1} - T_i = Q_T \tag{5.20}$$

where S_O is the change of storage of heat in the slab which is proportional to the rate of change of mean temperature. T_i and T_{i+1} are the net horizontal transports of heat into and out of the slab. Their difference is known as the heat flux divergence. Positive values indicate heat divergence and negative values indicate heat convergence. Equation 5.20 indicates that a net heat flux, Q_T, into the ocean must be accompanied by either a change in the heat storage or by the net advection of heat out of the slab. To obtain the heat transport across the boundary of each slab, it is assumed that the northward flux across the strip closest to the North Pole is zero. Hence, by summing consecutive strips, T_{i+1} can be obtained:

$$T_{i+1} = Q_T - S_O + T_i \tag{5.21}$$

The small contributions of frictional dissipation by tides and currents, by precipitation and by radioactive decay in the Earth's interior, have all been neglected. All of these factors are likely to contribute less than $1\,W\,m^{-2}$ to the net heat flux. In the ocean, and in marked contrast to over the land, the storage component, S_O, can be as large as the net heat flux, Q_T, during the seasonal cycle because of the important role played by the ocean in increasing the thermal inertia of the atmosphere–ocean system. Figure 5.3 shows a map of S_O, $T_{i+1} - T_i$ and Q_T for the Northern Hemisphere during the seasonal cycle. Comparison of S_O and Q_T shows the large accumulation of heat in the middle latitudes during the summer, and the release of that stored heat during the winter. However, it is noted that the middle-latitude heat loss in winter is greater than the heat accumulated in summer, and so there must be a poleward flux of heat to maintain the balance. This flux is, on average, about 2–3 petawatts (1 petawatt $=$ 10^{15} watts) at 25ºN.

The change of heat storage occurs mainly over the upper 100 m of the ocean, except in high latitudes of the Atlantic Ocean where it occurs over 500 m. It is also noted that, averaged over the seasonal cycle, the storage term tends to cancel out, and it is therefore reasonable to assume that, over a long period of observations, say 25 years, the change in storage will become small, and a balance between the transport of heat and the net heat flux across the surface will occur (equation 5.21). This is a steady-state assumption which implies that, in the long term, the mean ocean temperature is a constant. It is not possible to test this assumption in the ocean because of insufficient high-quality data over a long enough period of time. However, long-term trends in the sea surface temperature have shown increases of 0.5 K over the last 100 years. If this trend is occurring over the whole depth of the ocean, then an estimated net heat flux into the ocean of $2.5\,W\,m^{-2}$ is required. (This is likely to be an overestimate because the mixing time scale for the whole ocean is between 200 and 1000 years, and therefore this warming trend would not be shown by the whole mass of the ocean.) Compared with the net heat divergence shown in Figure 5.3, the steady-state assumption yields an error which is generally less than 10%.

Figure 5.4 shows the long-term mean meridional transport of heat in the ocean, estimated by applying equation 5.20 to each ocean basin. This map gives only a rough picture of the heat transport in the ocean

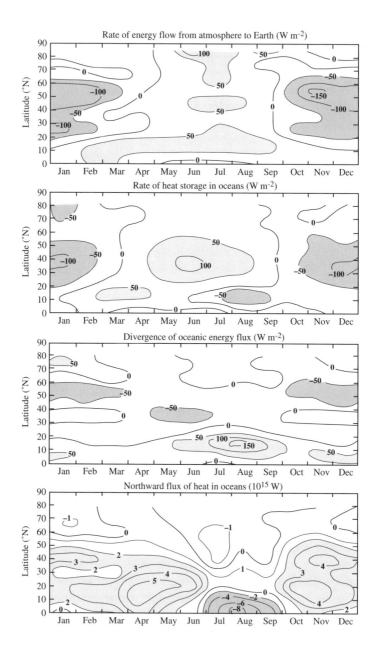

Figure 5.3   Vertical heat flux into the ocean, Q_T, storage, S_o, heat divergence and heat transport, T_i, in the Northern Hemisphere

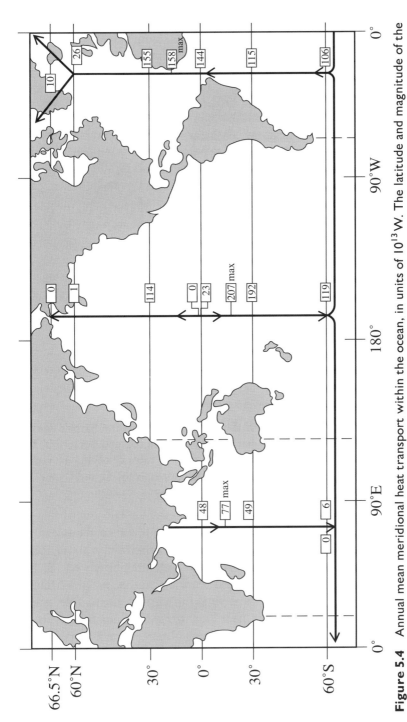

Figure 5.4 Annual mean meridional heat transport within the ocean, in units of 10^{13} W. The latitude and magnitude of the maximum meridional transport is shown for each ocean basin. Broken lines show meridians used as boundaries between oceans in high southern latitudes

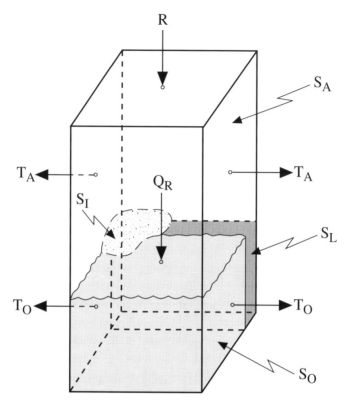

Figure 5.5 Schematic diagram of the different terms in the Earth's energy budget. R and Q_R are the net radiation fluxes at the top and bottom of the atmosphere, respectively. S_A, S_I, S_L and S_O are the rates of energy stored in the atmosphere, snow and ice, land, and ocean, respectively. T_A and T_O are the horizontal transports of energy in the atmosphere and the ocean

because of the errors in estimating the net heat flux into the ocean, as discussed in Section 5.1, and because of the uncertainty about long-term changes in storage. Of particular interest, however, is the difference between the Atlantic and Pacific Oceans. In the Atlantic Ocean, there is a northward heat flux throughout the whole ocean which is maintained by a net heat flux from the Indian and South Pacific Oceans. In contrast, in the Pacific Ocean, there is a poleward heat flux in both hemispheres with little or no net heat flux across the equator. It is also noted that the Arctic Ocean has a substantial heat flux from the North Atlantic Ocean, but gains no significant heat from the Pacific Ocean through the Bering Straits.

An overall conservation equation for the entire atmosphere–ocean can be determined in a similar way to equation 5.20. A box extending from the base of the ocean to the top of the atmosphere is shown in Figure 5.5. The net heat flux, R, into the top of the atmosphere is simply the sum of the net solar radiation and net outgoing planetary radiation, which can be determined from satellite observations. The divergence of atmospheric energy, i.e. internal energy, potential energy and latent heat flux, is represented by T_A, the atmospheric storage term by S_A, the land storage and ice storage by S_L and S_I, respectively, the ocean storage by S_O and the divergence of the ocean by T_O, then:

$$R = T_A + T_O + S_A + S_O + S_L + S_I \qquad (5.22)$$

If, over a long period of time, the storage terms are assumed to be negligible, then:

$$R = T_A + T_O \qquad (5.23)$$

The term $T_A + T_O$ is the total energy divergence of the atmosphere–ocean system. In a similar process to the determination of the poleward transport of heat in the ocean, equation 5.23 can be used to determine the total poleward transport by atmosphere and ocean. Figure 1.6 shows that poleward of 40° there is a net deficit of radiation, because the outgoing long-wave radiation exceeds the incoming solar radiation, and so a poleward transport of energy is required to maintain the balance. The atmospheric energy transport can be obtained by subtracting the ocean transport from the total transport (Figure 5.6). These studies have shown that the atmosphere and ocean carry similar quantities of heat from low latitudes to high latitudes over an annual period. The ocean heat transport is a maximum at 20° latitude, whilst the atmosphere transport reaches a maximum at about 40° latitude.

5.3 The measurement of the water budget

The vertical flux of water, W, across a horizontal surface is given by:

$$W = P - E \qquad (5.24)$$

where P is the precipitation, including dewfall, and E is the evaporation. Both quantities are usually expressed in units of millimetres per year or centimetres per year. The precipitation is measured almost entirely

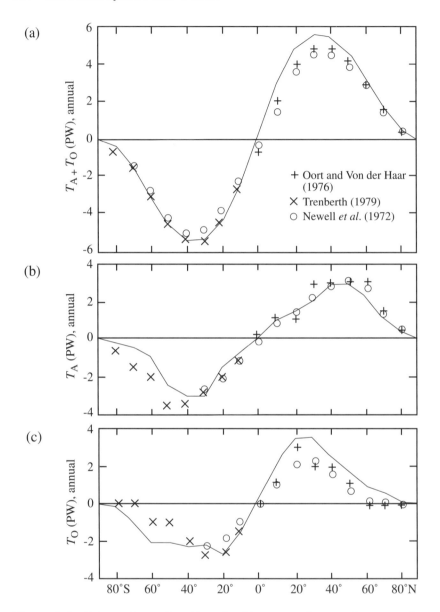

Figure 5.6 Poleward transfer of heat by: (a) ocean and atmosphere together $(T_A + T_O)$, (b) atmosphere alone (T_A), and (c) ocean alone (T_O). The total heat transfer (a) is derived from satellite measurements at the top of the atmosphere, that of the atmosphere alone (b) is obtained from measurements of the atmosphere, and (c) is calculated as the difference between (a) and (b). 1 petawatt (PW) = 10^{15} W

from land-based measurements using rain gauges. At a standard climatological station, when care is used in siting the rain gauge, the precipitation can be determined to an accuracy of approximately 1–2%. However, over the oceans, precipitation measurements using rain gauges are neither as frequent nor as accurate as over the land. The reasons for this are related to the problem of wind flow around the ship carrying the gauge and, to a lesser extent, the problem of spray. Comparisons between rain gauges mounted on different parts of a ship have shown that differences of 50–100% can be expected, and a number of indirect techniques have, therefore, been employed to measure the rainfall over the ocean. One technique uses the visual observations of rain intensity, which are recorded regularly by merchant and weather ships as part of their meteorological reports, and converts them into numerical values of rainfall intensity. As an example, light continuous rain would be assigned a rainfall intensity of 0.6 mm hour^{-1}, whilst moderate continuous rain would be given a value of 1.9 mm hour^{-1}. By this method, annual mean distributions of rainfall can be obtained to an accuracy of about 10%. This technique is probably best suited to higher latitudes where the rainfall is less intense, but more frequent, than in the tropics. Another indirect method is to use radar to measure the back-scattered radiation from large water droplets. The radar can be calibrated by comparing simultaneous measurements of back-scattered radiation with rainfall rates determined by a conventional, land-based, rain-gauge network. This radar technique is not very accurate because the back-scattered radiation is dependent on the size distribution of the droplets. It will not be reliable when there is a large component of frozen precipitation. It is most useful for the detection of intense mesoscale rainfall systems and is limited to a range of about 100 km over land. Over the sea there is a reduced horizon, because of the limited height at which the radar is fixed onto the ship.

In recent decades there has been considerable interest in the use of satellites to measure precipitation. One technique measures the temperature of the cloud tops using infra-red sensors. This can be used to obtain the depth of a cloud which can, in turn, be empirically related to the intensity of precipitation from the cloud. This method is most successful for deep convective clouds and is, therefore, best suited to determine the rainfall distribution over the tropical ocean. However, as with all satellite techniques, its principal value lies in the determination of the spatial distribution of precipitation rather than for obtaining accurate point values. A second satellite method uses a microwave radiometer to measure the small amounts of microwave radiation emitted by the suspended liquid water in the atmosphere. Changes in the liquid water content can be related to the precipitation rate, and thus an instantaneous picture of the precipitation pattern may be obtained. Both satellite techniques are limited by the frequency of satellite passes over a particular region; perhaps only one to three passes per day. Rainfall is highly variable in both time and space, and so these techniques require further development if they are to be of use in obtaining quantitative climatological precipitation over the globe.

The direct measurement of evaporation from a ship is no less difficult than that of precipitation. The simplest method is to measure the rate of evaporation from a pan filled with salt water and suspended in gymbals. An accurate determination of the water lost can be made by measuring the change in salinity. In practice, this method is unreliable because the evaporation from a pan at deck level will generally be higher than that at sea level due to the vertical gradient of humidity and wind speed over the ocean. To be representative of the evaporation from the ocean surface, the pan would have to be suspended or floated on the surface, and this is only possible in the smoothest of seas. Secondly, solar heating of the pan itself would result in anomalously high evaporation rates. Comparisons of the evaporation at the sea surface with that at deck level have indicated systematic errors of about 50%. Because of all of the limitations and difficulties discussed above, the indirect techniques, such as the bulk aerodynamic method discussed in Section 5.1, are found to have smaller systematic errors, of about 10%, than the pan method. These indirect techniques can be used to make routine calculations from merchant ships' reports. It is noted that an evaporative heat flux of $100 \, \text{W} \, \text{m}^{-2}$, such as observed at the equator, is equivalent to an evaporation of $126 \, \text{cm} \, \text{year}^{-1}$.

5.4   Observations of the water budget

For the reasons discussed above, the distribution of precipitation is only accurately known over land areas. In the Pacific Ocean, the large number of islands in the western and central regions allow a good estimate of the rainfall over the ocean to be obtained, although it is important to realize that islands, especially if they have hills or mountains, will tend to be preferential sites for tropical convection and precipitation. In the tropical South Atlantic, the paucity of island sites results in a large uncertainty in the rainfall distribution. Between the equator and 40ºS, there are only seven islands, not all of which record rainfall. However, despite the uncertainty in quantitative measurements over the ocean, the general global distribution is probably correct and, in recent years, the previously described satellite techniques have enabled maps of the major precipitation belts to be obtained.

The most significant feature of the precipitation distribution in tropical latitudes is the relationship between the sea surface temperature and the rainfall belts. High rainfall in both the Pacific and Atlantic Oceans occurs along the ITCZ, which corresponds closely to the thermal equator.

Of particular note is the very high rainfall over the equatorial regions of South America and Africa, and also the maritime continent of Indonesia, which exceeds 300 cm year^{-1}. By contrast, in parts of the eastern tropical and sub-tropical ocean, the rainfall is less than 10 cm year^{-1}. This dramatic decrease is particularly noticeable in the south equatorial Pacific, where the Galapagos Islands have a desert climate, whilst only 300 km northwards, in the ITCZ, the rainfall may exceed 300 cm year^{-1}. The oceanic desert regions are associated with lower sea surface temperatures which, in turn, are related to the upwelling of cold sub-surface water and the equatorial advection of colder water by the sub-tropical anticyclonic gyres.

With the use of climatological cloud distributions derived from satellite photographs, convergence zones and regions of higher precipitation have been located in all three ocean basins. One of the most significant is the South Pacific convergence zone which extends south-eastwards from Indonesia into the central South Pacific. Similar cloud zones have been observed over the South Atlantic extending from the Amazon basin, and from East Africa into the southern Indian Ocean. In higher latitudes, the precipitation rates are much lower due to the lower water-holding capacity of the cooler atmosphere. The evaporation rate, which is related to both the vertical gradient of humidity and to the wind speed, is large over the tropical trade-wind zones where both the

Figure 5.7 Global distribution of evaporation minus precipitation $(E - P)$ over the ocean. Units are cm year^{-1}; solid lines indicate regions where $E > P$, dotted lines where $E < P$.

wind speed and the vertical gradients of humidity are large. In these regions, evaporation reaches 150–200 cm year^{-1}. In the middle latitudes, large evaporative fluxes of up to 250 cm year^{-1} are seen over the western North Pacific and western North Atlantic due to advection of dry continental air over a warm ocean surface.

Over the whole globe evaporation must, of course, be in balance with precipitation. The global average precipitation required to balance evaporation is about 103 cm year^{-1}, and a horizontal transfer of water must occur to balance regions of water deficit with regions of water surplus. Figure 5.7 shows the latitudinal variation of the components of the water balance. A net surplus of water (i.e. $W > 0$) exists over the equatorial region, associated with the ITCZ, and polewards of 40° latitude, associated with the middle-latitude cyclones. In the tropical and sub-tropical zones there is a net deficit (i.e. $W < 0$) as a result of the low precipitation over the ocean and desert regions, and the high evaporation in the trade-wind zones. To maintain this latitudinal variation of W, it is necessary to transfer fresher water in the ocean from the higher latitudes and the equatorial zone into the sub-tropical regions. In the atmosphere, a transfer of water vapour in the opposite direction is required to maintain the regions where precipitation exceeds evaporation. Most of the water vapour is contained in the lower 5 km of the atmosphere and these water-vapour transfers must, therefore, occur in the lower layers of the atmosphere. The trade winds, which occupy the lowest 2–3 km of the atmosphere, are responsible for the transfer of water vapour from the sub-tropical ocean into the ITCZ. During the Indian monsoon, the low-level south-west monsoon winds transfer water vapour from the southern Indian Ocean and the Arabian Sea onto the Indian sub-continent. At higher latitudes, the warm moist poleward winds, associated with the warm sectors of extra-tropical cyclones, transfer water vapour into regions of precipitation surplus.

Over the continents, the precipitation generally exceeds the evaporation by about 25 cm year^{-1}, whilst over the oceans, evaporation exceeds precipitation. The continental surplus is transferred by rivers into the ocean to balance the ocean freshwater deficit. The largest freshwater surplus per unit area occurs over South America, and the lowest over Australia. These continental surpluses must be balanced by a net flux of atmospheric water vapour from the ocean to the continents.

Figure 5.8 An estimate of the transfer of fresh water ($\times 10^9\,\mathrm{kg\,s^{-1}}$) in the world's ocean. In polar and equatorial regions, precipitation and river runoff exceed evaporation, and hence there is an excess of fresh water, while in subtropical regions there is a water deficit. A horizontal transfer of fresh water is therefore required from regions of surplus to regions of deficit. F_P and F_A refer to the freshwater fluxes of the Pacific–Indian throughflow and of the Antarctic Circumpolar Current in the Drake Passage, respectively

Although evaporation exceeds precipitation over the global ocean, there are significant differences between individual ocean basins. In the Atlantic and Indian Oceans, evaporation exceeds precipitation, whilst in the Pacific and Arctic Oceans, precipitation dominates evaporation. For the balance to be maintained, a net transfer of fresh water from the Pacific and Arctic Oceans into the Atlantic and Indian Oceans is required (Figure 5.8). The water budget for the Atlantic Ocean shows that about 75% of the deficit is supplied by the surrounding continents as runoff, and the remainder by freshwater transport from the Arctic and Pacific Oceans.

In conclusion, it has been shown that the surface water budget allows a picture of the global transfers of atmospheric water vapour and of fresh water in the ocean to be ascertained. As yet, this picture is still crude because of the inaccuracies of the precipitation and evaporation estimates but, with improved satellite techniques, it should be possible to derive water budgets for seasonal and interannual variations, as well as for the long-term means.

5.5   The salt budget of the ocean

The total mass of the ocean is 1.4×10^{21} kg and, if a global average salinity of 34.7 psu is assumed, then the total mass of dissolved salt is 4.8×10^{19} kg. It has been estimated that the rate of salt input into the ocean is about 3×10^{12} kg year^{-1}, and therefore the replacement time for the ocean salt is approximately 16 million years. This simple calculation illustrates the point that the sources and sinks of salt in the ocean are only of concern on a geological time scale. For ocean circulation problems, a time scale of less than 1000 years is applicable, and therefore it is a good approximation to assume that the total salt content of the ocean is constant. This has the important consequence that observed salinity variations in the ocean are the result of dilution and concentration brought about by imbalances in the surface water budget.

Over the ocean basins, away from major river discharges, the salinity distribution is well correlated to the net surface water flux, W. The principal features are that the largest surface salinities occur in the sub-tropical zone where there is a net deficit of water, and that the lowest surface salinities occur in high latitudes where precipitation exceeds evaporation. A minimum is also apparent in the region of the ITCZ. As might have been anticipated from the discussion in the previous section, there are considerable differences between the various ocean basins. The North Pacific Ocean has an average surface salinity of about 34.2 psu, whilst the North Atlantic Ocean has a salinity of 35.5 psu. These observations are related to the surplus of fresh water in the Pacific Ocean and the deficit in the Atlantic Ocean. Furthermore, the low salinity in the high latitudes of the North Pacific maintains a stable density stratification, known as a halocline, which inhibits the formation of deep water. The formation and stability of the halocline in the Arctic Ocean is related partly to the excess of precipitation over evaporation, and partly to the influx of fresh water from runoff, particularly by Siberian rivers. A major concern has been that a reduction of runoff caused by a redirection of Siberian rivers from the Arctic Ocean to the Caspian Sea may affect the stability of the Arctic halocline. It has been argued that weakening of the halocline would enable deep convection to occur which would, in turn, bring warmer water to the surface and inhibit the development of sea ice. Although this is only a possibility, it does demonstrate the point that small variations in the water balance may produce dramatic changes in regional climate.

A useful example of the application of the principle of the conservation of salt may be found in the Red Sea. There is a surface flow of low-salinity water into the Red Sea through the Strait of Bab el Mandeb and an outflow of high-salinity water over a sill at a depth of 110 m. If F_I is the volume flux of water into the Red Sea, with a salinity of S_I, and F_O is the volume flux of water out of the Red Sea, with a salinity S_O, then for the conservation of salt across the Straits of Bab el Mandeb:

$$\rho_I F_I S_I = \rho_O F_O S_O \qquad (5.25)$$

It can be assumed that $\rho_I \simeq \rho_O$ and for the conservation of water:

$$F_I = F_O - W - R_O \qquad (5.26)$$

where W is the net water flux through the surface of the Red Sea and R_O is the runoff. Thus, eliminating F_O:

$$F_I S_I = (F_I + W + R_O)S_O$$

$$\therefore \quad F_I\left[\frac{S_O - S_I}{S_O}\right] = -[W + R_O] \qquad (5.27)$$

The salinity, S_O, is 40‰ and S_I is 37‰. Measurements across the Straits suggest $F_I = 0.5 \times 10^6 \, \text{m}^3 \, \text{s}^{-1}$, and therefore the net surface flux of water is approximately $-3.75 \times 10^4 \, \text{m}^3 \, \text{s}^{-1}$. The negative sign indicates an excess of evaporation over precipitation and runoff. The surface area of the Red Sea is $438 \times 10^9 \, \text{m}^2$, and therefore the excess evaporation over precipitation and runoff is $2.70 \, \text{m year}^{-1}$. Since precipitation is generally less than 20 cm year^{-1} and the runoff term is negligible, the evaporation rate is close to this figure.

Equation 5.27 is generally applicable to any semi-enclosed sea or estuary where the vertical circulation can be represented by a two-layer fluid (Figure 5.9). Some conclusions can be drawn from this equation. When evaporation exceeds precipitation and runoff, $(W + R_O) < 0$ and $S_O > S_I$, i.e. the outflow will be more saline than the inflow. If the

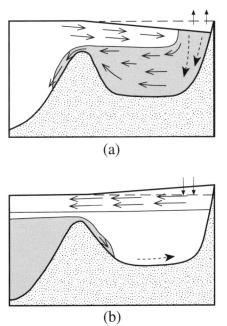

(a)

(b)

Figure 5.9 The two principal types of sill basin. (a) Evaporation > precipitation + runoff. (b) Evaporation < precipitation + runoff temperature changes are much smaller than the salinity changes, then the more-saline water will sink and flow out underneath the less-saline inflow. This situation is typical of the Red Sea, the Persian Gulf and the Mediterranean Sea, all of which produce saline bottom waters. In the northern extremities of the Red Sea and in the Persian Gulf, salinities as high as 42 psu occur, whilst in the Mediterranean Sea, salinities of 38 psu are typical. In higher latitudes, precipitation and runoff exceed evaporation, i.e. $W + R_O > 0$, and therefore $S_O < S_I$. Outflow will thus be less saline than inflow and this fresher water will flow out of the surface layer, whilst the more-saline ocean water will flow in at depth. This situation is representative of the Baltic Sea, where surface salinities as low as 10 psu can occur. From equation 5.27 it can also be seen that the salinity difference, $S_O - S_I$, is proportional to the ratio of $(W + R_O)/F_I$ and, if the volume flux into the basin is considerably larger than the net water flux, $W + R_O$, then the salinity difference between the inflow and outflow will be small. On the other hand, large salinity changes will occur if the volume flux is small relative to the net water flux. The Baltic Sea, for instance, has a very shallow sill depth of 18 m, the volume flux is very small compared with the freshwater input, and hence the surface water has a low salinity.

The value of equation 5.27 lies in obtaining either a consistency check on the salt budget of an estuary or semi-enclosed sea, or in determining the value of one of the terms in the equation such as precipitation which, as previously discussed, is difficult to measure directly. The principle can also be applied to the salt balances of ocean basins such as the Arctic Ocean. In this case, the situation is complicated by the large number of inflows and outflows, and by the fact that the vertical circulation may be rather more complicated than that assumed for the Red Sea. In ocean basins such as the Atlantic Ocean, direct measurements of the volume fluxes of the water masses are only possible in limited areas, but equations based on the conservation of salt, mass and heat have been solved, using inverse analysis, to provide large-scale transports.

5.6   Temperature and salinity relations in the ocean

In the previous section it was shown how the surface water flux, runoff and circulation determine the observed salinity, and it is reasonable to acknowledge that the surface heat budget, in conjunction with circulation, can determine the observed temperature of the ocean. It is clear that the relationship between temperature and salinity will give useful information about the sources of water masses, and about the circulation and mixing rates in the ocean. Close to the surface, the temperature and salinity characteristics will be constantly changing in response to synoptic and seasonal variations in heat fluxes and water fluxes. However, below the depth of seasonal influence, on average about 100 m, the temperature and salinity of a water parcel will only change by mixing with the surrounding water masses, and therefore it can be said that the temperature–salinity (T–S) properties of the parcel will be conserved. In the deep ocean, potential temperature, θ, is used rather than the actual temperature, T, because of the adiabatic compression.

Figure 5.10 shows the temperature and salinity profiles taken at two oceanographic stations in the Gulf Stream about 200 km apart. At first glance there is no obvious similarity but, when the information is plotted on a T–S diagram, also shown in Figure 5.10, it can be seen that they have an identical characteristic shape. In this case, the water mass at one station has been vertically displaced by 600 m relative to the other station. This example shows that a T–S relationship will allow a particular water mass to be labelled and to be followed in the ocean. In oceanography, a point on a T–S diagram is referred to as a water type, whilst a straight-line relationship is known as a water mass. Thus, the water type refers to the original T–S during formation at the surface, whilst the water mass evolves by mixing between different water types. Figure 5.11a shows the progressive mixing of two water types, A and B. In order to conserve temperature and salinity, mixtures of the two water types must lie on the line AB, and the position on the straight line is determined by the relative concentration of each water type. Hence, a succession of different water types are formed. The ensemble of water types is known as a water mass. Figure 5.11b shows the mixing of three water types, not dissimilar to those in the Atlantic Ocean. The mixed water will be located within the triangle ABC and its actual position will depend on the relative proportions of the three types. If the rate of mixing between the water mass above and the water mass below is equal, then the salinity minimum will be located along the line BD. An example of the progressive mixing of water types is the distribution of the Antarctic Intermediate Water in the

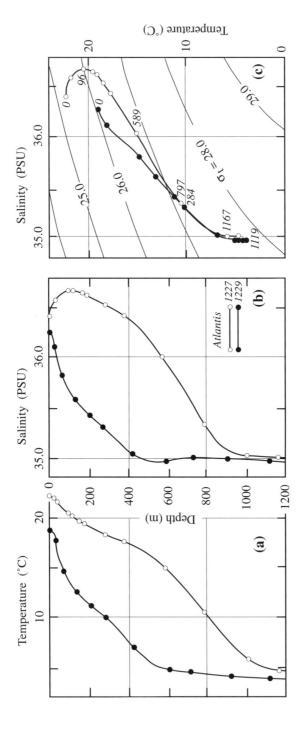

Figure 5.10 Relation between temperature and salinity for two neighbouring *Atlantis* stations 1227 and 1229 in the Gulf Stream east of Cape Hatteras. (a) Vertical temperature distribution (°C). (b) Vertical salinity distribution (psu). (c) T–S diagram, with lines of constant density (σ_t)

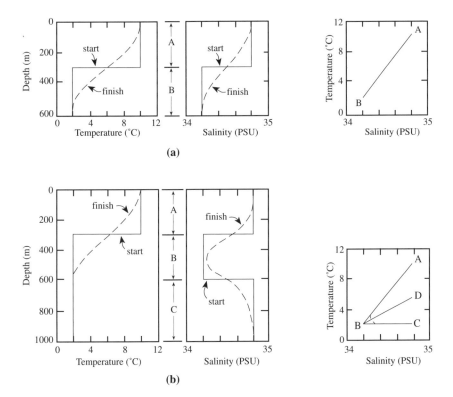

Figure 5.11 Temperature–salinity relationships resulting in the mixing of water masses. (a) Mixing of two water types. On the T–S diagram, mixing of the two water types is shown by the straight line connecting the T–S points characteristic of the two original water types. (b) Mixing of three water types. On the T–S diagram, the mixing process is shown by two straight lines, indicating that water types A and B mix, and that water types B and C mix. The lines meet at a point representing water type B, now altered by mixing with water above and below so that its original identity has been lost and it can no longer be represented by a single point on the T–S diagram. For equal mixing above and below, the mixture will lie along the line BD

Atlantic Ocean, as shown in Figure 5.12. This water type is found in the Antarctic convergence zone at about 50°S, where it has a temperature of 2.2°C and a salinity of 33.8 psu. It sinks to a depth between 700 and 1500 m, and then it progresses northwards as a tongue of low-salinity water. It can be identified on the T–S diagram of the Atlantic Ocean, shown in Figure 5.12, as a salinity minimum. Successive T–S diagrams show that the water mass becomes warmer

Figure 5.12 Variations with latitude of the salinity minimum characterizing the Antarctic Intermediate Water (AIW) between 17°S and 9°N in the Atlantic. The AIW is sandwiched between warmer, saline Central waters above and saline North Atlantic Deep Water below. These water masses mix with the AIW as it progresses northward

and more saline as it mixes with the Central Atlantic water above and the Atlantic Deep Water below. Eventually, at about 20°N, it can no longer be identified on a T–S diagram. From the T–S diagrams, the rate of mixing of the Antarctic Intermediate Water can be deduced, and thus an estimate of the volume of the original water type can be made.

Figure 5.13 shows a three-dimensional diagram representing a census of the world's water masses by Worthington (1981). The vertical co-ordinate is the volume of water having a mean temperature in the range ±0.05°C and a mean salinity in the range ±0.005 psu. Thus, the

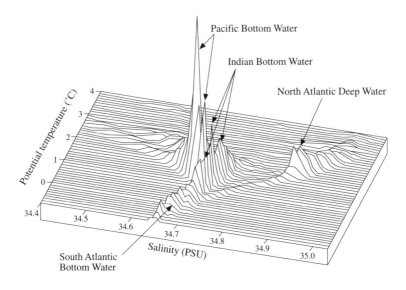

Figure 5.13 Simulated three-dimensional θ–S diagram of the water masses of the world ocean. The potential temperature (θ) is used rather than *in situ* temperature (T), in order to remove the adiabatic effect

peak in Figure 5.13 corresponds to a potential temperature between 1.1 and 1.2°C, and a salinity between 34.68 and 34.69 psu, and has a volume of $26 \times 10^{15}\,\mathrm{m}^3$. This water mass has a larger volume than all of the water above 19°C and larger than the total volume of all the continental ice caps. The census shows that the major water masses of the ocean have a relatively small range of temperature and salinity, distributed about a global mean temperature of 3.51°C and a mean salinity of 34.72 psu. The much warmer surface waters and the saline deep waters of the Mediterranean and Red Seas are insignificant compared with the deep polar waters which fill all of the ocean basins. Thus, the major water masses of the ocean are probably not significantly different to those occurring during the Ice Ages and the amelioration of the climate by the ocean is limited to the effect of a rather shallow, warm surface layer. The most uniform water masses are found in the North Pacific Ocean where there are no sources of bottom water. All these deep-water masses have their origin in the Atlantic and Antarctic Oceans, and therefore the mixing process has had sufficient time to homogenize the deep-water masses in the Pacific. The Indian Ocean tends to have a relatively uniform salinity of 34.70 psu, but a larger range of temperatures. The Atlantic Ocean is quite different from either the Indian or the Pacific Oceans. It has a large number of water masses ranging from -1 to $+4^\circ C$, and most are relatively saline compared with the Indian and Pacific Oceans. Of note is the formation of North Atlantic Deep Water in the Labrador and Greenland Seas during late winter, and the formation of Weddell Sea Bottom Water which, at $27.88\ \sigma_t$, is the densest water mass in the ocean, having a temperature of $-0.5^\circ C$ and salinity of 34.66 psu. Dense water is also found at the bottom of the Norwegian Sea, the Red Sea and the Mediterranean Sea but, because of the sills at the entrances to these seas, these dense waters cannot flow freely into the open ocean.

A novel application of the T–S relationship was made by Stommel and Csanady (1980) to estimate the ratio of the freshwater flux to the heat flux. By analogy with equation 5.5, the gradient of salinity is proportional to the freshwater flux and the gradient of temperature to the heat flux. By taking the ratio of the temperature gradient to the salinity gradient, Stommel and Csanady obtained the ratio of the freshwater flux to the heat flux.

The conductivity, temperature and depth recording system (CTD) is the main method for determining the T–S properties of the world's oceans. CTD systems consist of three sensors whose resistances vary with the temperature, conductivity and pressure of their surroundings. The resistances of the three sensors are measured by an audio-oscillator whose frequency is determined by the resistance. The signals are sent up a cable to be processed on board the research ship. The conductivity of the sea water is measured using platinum electrodes, and it is dependent on both temperature and salinity. Thus, a temperature correction to the conductivity signal is required to obtain accurate salinity values. Usually, two temperature sensors are employed. One is a thermistor, which has a very fast response time, and the other is a platinum resistance thermometer, which is very accurate. The thermistor measures small variations in temperature, and the platinum resistance thermometer is used in the absolute calibration of the thermistor and the conductivity sensor. By this method, *in situ* temperatures can be measured to $\pm 0.001^\circ C$, salinity to ± 0.005 psu and depth to an accuracy of 1 m. Additional probes can be attached to measure optical and chemical properties in the water column. Thus, from one cast of a CTD system, a high-resolution vertical distribution of temperature and salinity can be obtained.

These measurements have shown that the T–S relationships for water masses are not as uniform as shown by classical methods. Distinct layers of uniform water have been found to be separated by rapid changes in temperature and salinity. These uniform layers range from as little as 1 m in depth up to 100 m. The small-scale variations in T–S relationships are known as the fine structure, and it is now believed that these fine structures are direct evidence for water-mass transformation in the deep ocean. One process which can cause this fine structure at the boundaries of water masses is double-diffusive convection. One example is salt fingering, which depends on the much larger diffusivity of heat compared with the molecular diffusivity of salt. Consider two layers of water, the upper one being salty and warm, whilst the lower layer is cold and fresh. If it is assumed that temperature dominates the density of each layer, the two fluids will be hydrostatically stable. Consider now a parcel of fluid moved from the upper layer to the lower layer where, because of the larger heat diffusivity, it will lose heat more rapidly than it will lose salt. The parcel will lose its buoyancy, will become statically unstable and sink. A parcel moving from the lower layer to the upper layer will gain heat more rapidly than salt, will become buoyant with respect to its surroundings and accelerate upwards. Hence, alternating upward (fresh)

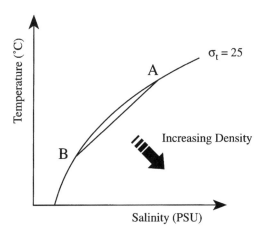

Figure 5.14 Mixing of two water masses (A, B) having similar densities (σ_t = 25). Any mixture will fall on the straight line AB and therefore will always have a density greater than σ_t = 25

and downward (salty) motion is produced, which transfers salt downwards across the interface. The release of convective energy by the salt fingers results in the formation of well-mixed layers above and below the interface. Such fine structure has been found under the Mediterranean outflow where salty and relatively warm water from the Mediterranean Sea overlies the less saline, but colder, North Atlantic Deep Water.

A second process, which can lead to mixing between water masses, depends on the non-linear equation of state for sea water, and it is referred to as 'caballing' in the oceanographic literature. Consider two parcels of fluid having different temperatures and salinities but equal densities, as depicted in Figure 5.14. The resultant mixture will lie along the line AB, its position being determined by the relative proportions of the two original fluids. However, because of the curvature of the lines of constant density known as isopycnals, the mixture will be more dense than the original water parcels and therefore the mixture will sink. The curvature of the isopycnals is a maximum at low temperatures and, therefore, caballing is particularly important in Antarctic and Arctic waters. The formation of Antarctic Bottom Water has been attributed to this process. In the Weddell Sea, the surface water is generally colder and less saline than the underlying Antarctic Circumpolar Water. In winter, the formation of sea ice releases extra brine into the surface layer which, in conjunction with winter cooling, may produce a water mass of similar density to the underlying one. In this case, mixing of the upper and lower water masses may produce a mixture which has a density greater than the deep water, and which will therefore sink.

The above fine-structure processes are not the only ones by which water masses are diluted by mixing. In the case of outflows of dense water from semi-enclosed seas, the plumes may be rapidly mixed by entrainment with the surrounding water. Occasional intense gravity currents are observed over the Faeroes–Iceland ridge and in the Faeroes Channel, when dense Norwegian Sea water flows over the sills into the Atlantic Ocean. The plume accelerates down the slope because of the density differences with the ambient water and may reach speeds of $1 \, \mathrm{m \, s^{-1}}$. Intense mixing of the plume occurs, which increases its volume of water and reduces its density, to form North Atlantic Deep Water. Calculations of the mixing ratio have shown that the overflow water is composed of $1 \times 10^6 \, \mathrm{m^3 \, s^{-1}}$ Norwegian Sea water and $3 \times 10^6 \, \mathrm{m^3 \, s^{-1}}$

entrained Atlantic water. The dense high-salinity plume from the Red Sea also experiences mixing with Indian Ocean water as it moves off the Bab el Mandeb sill into the Gulf of Aden. Here, the plume descends to a depth of 1000 m before it reaches a similar density to the surrounding water and spreads horizontally into the Indian Ocean.

A second important horizontal mixing process is caused by energetic mesoscale eddies, typically 50–200 km in diameter. These eddies have been found in all parts of the ocean and can homogenize water masses, especially in intense frontal zones such as the Gulf Stream and the Antarctic convergence zones. It is thought that the diffusion of Mediterranean water into the North Atlantic is primarily due to horizontal mixing by eddies. In 1978, a water mass having a temperature and salinity characteristic of Mediterranean water was located in the western Atlantic in a small intense mesoscale circulation. It had apparently taken about three years to cross the Atlantic Ocean and in that time it had not mixed, to any significant extent, with its environment. This observation suggests that there is still much to be understood about the mixing of water masses in the ocean.

5.7   Tracers in the ocean

In the early 1970s, a research programme known as GEOSECS (Geochemical Ocean Sections Study) was initiated to measure the distribution of natural and anthropogenic tracers in the world ocean. Although oceanographers had used naturally occurring tracers, such as dissolved oxygen and carbon-14, to study water masses before this experiment, GEOSECS was the first comprehensive attempt to measure these tracers, and many others, on a global basis.

Table 5.2 shows tracers which have been successfully used in the ocean to determine water-mass production, mixing and circulation. Dissolved oxygen has been measured in the ocean, in conjunction with salinity and temperature, since the 1930s, and much information on its global distribution is available. Oxygen is absorbed most efficiently at low temperatures, and thus the largest concentrations are found in Antarctic and Arctic waters. It is a good tracer for studying the mixing of water masses, such as the North Atlantic Deep Water and Antarctic Bottom Water, but it has to be monitored continuously because it is consumed by bacteria and animals even in the deep ocean.

Table 5.2 Selection of artificial and natural tracers in the ocean

Tracer	Origin	Conservative	Half-life (years)	Comment
Dissolved O_2	Surface exchange	No	–	Consumed by bacteria and zooplankton
Silica (SiO_2)	Bottom sediments	No	–	Removed by upper ocean
Nitrate (NO_3)	Bacteria and zooplankton	No	–	–
Phosphate (PO_4)	Bacteria and zooplankton	No	–	–
$[O_2] + 9[NO_3]$	–	Quasi-conservative	–	Not conserved in low O_2 water
$[O_2] + 135[PO_4]$	–	Quasi-conservative	–	–
^3H	Cosmic rays Nuclear testing	Conservative	12.4	Natural tritium very small compared with bomb tritium
^{90}Sr	Nuclear testing	Conservative	28.6	Concentrated in tropical waters Recorded in growth rings of coral
^{14}C	Cosmic rays Nuclear testing	Conservative	5730	Appears in solution and particulates
^{137}Cs	Nuclear testing	Conservative	30.2	Requires large samples for measurement
^{39}Ar	Cosmic	Conservative	250	Requires very large samples
Chlorofluorocarbons (CFC or freons)	Anthropogenic	Conservative	–	Sophisticated analysis required as contamination is a hazard

Nitrates and phosphates are excreted by animals and bacteria, and they have been used as short-term tracers.

Silica is produced from siliceous bottom sediments such as red clays, which are plentiful in the Pacific Ocean and, to a lesser extent, in the Indian and Atlantic Oceans. Silica is consumed by biological processes in the upper layers of the ocean and it is, therefore, most useful for studying bottom and intermediate water circulations. In the North Atlantic Deep Water, the silica concentrations are low compared with those in the Antarctic Circumpolar Waters and Antarctic Bottom Waters, and it is therefore an ideal tracer when studying the mixing of North Atlantic Deep Water into Antarctic Circumpolar Current. North Atlantic Deep Water is composed of a number of different water types and so the θ–S characteristics vary accordingly. However, the silica concentration is remarkably uniform, and it is therefore a more consistent tracer than the θ–S profile. The North Atlantic Deep Water (low silica concentrations) flows south against the western boundary and the Antarctic Bottom Water (high silica concentration) flows northwards along the bottom. It is noted that in bottom waters, where the θ–S properties usually show only a small variation, the use of silica as a tracer is particularly useful.

The information obtained from the above natural tracers is only qualitative because of the lack of knowledge about rates of production and consumption. Quantitative information on the age and circulation time scale of any given water mass can be obtained using radioactive tracers. As Table 5.2 shows, some tracers are produced naturally from cosmic rays, such as ^{39}Ar, some are by-products of nuclear tests, such as ^{90}Sr and ^{137}Cs, and others, such as tritium, ^{3}H and ^{14}C, are by-products of both nuclear tests and cosmic rays. The long-lived, naturally produced isotopes are particularly useful because they are generally in a steady-state balance in the ocean, and therefore they can be used to estimate the age of the ocean waters and the ventilation time of the ocean. ^{14}C measurements, obtained before nuclear bomb tests significantly altered the ^{14}C distribution in the ocean, enabled the first estimate of the ventilation time, T, to be obtained thus:

$$T = V/R_v$$

where V is the volume of the ocean and R_v is the ventilation rate for deep water. For radiocarbon, the ventilation time is also given by:

$$T = \left[\frac{(^{14}C/C)_S}{(^{14}C/C)_D} - 1 \right] T(^{14}C) \qquad (5.28)$$

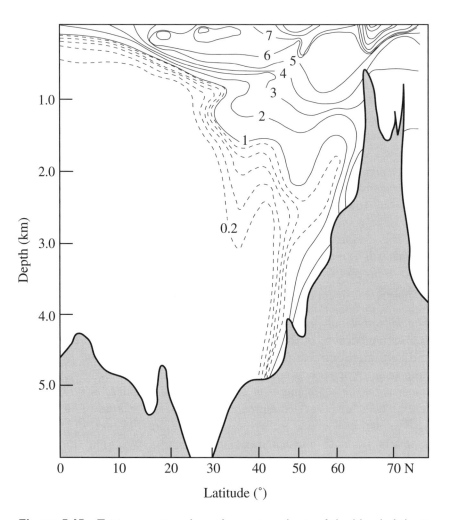

Figure 5.15 Tritium section along the western basin of the North Atlantic obtained during the GEOSECS programme. The tritium was introduced into the ocean from atmospheric testing of nuclear weapons in the 1950s and early 1960s. After the test-ban treaty, no further tritium was introduced into the ocean, and therefore it provides a unique marker of water masses in contact with the atmosphere during this period. Values are in Tritium Units (TU); 1 TU = 1 × 10^{-18} atoms of tritium per atom of hydrogen

where $(^{14}C/C)_S$ and $(^{14}C/C)_D$ are the carbon isotope ratios for the surface and deep waters, respectively, and $T(^{14}C)$ is the mean half-life of ^{14}C, which is 8200 years.

Equation 5.28 gives a mean ventilation time of 1400 years $\pm$ 250 years for deep-water circulation. Recent evidence has suggested that this is an overestimate, because it is often not possible to distinguish old water masses which may have been recirculated around the ocean a number of times from newly formed water masses. Measurements on North Atlantic Deep Water have indicated an apparent age of between 80 and 225 years. Argon is also a potentially useful natural radioactive tracer because it has a half-life similar to the ventilation time. However, at present it is technically difficult to determine because several cubic metres are required to obtain just one measurement.

The radioactive tracers produced from atmospheric nuclear-weapon tests between 1945 and 1963 have inadvertently provided a number of useful transient tracers. These tracers are not in equilibrium with the ocean and they cannot provide information on the age of water masses, but they have proved valuable for determining the progress of recently formed bottom and intermediate waters. The bomb isotopes best studied in the GEOSECS experiment were ^{14}C and ^{3}H. Anthropogenic carbon-14 can be distinguished from natural carbon-14 because of its very large signal in the surface layer. The distribution of tritium is not known prior to the bomb tests, but it is believed that the natural concentrations are much lower than the anthropogenic ones. ^{14}C and ^{3}H were released into the Northern Hemisphere's stratosphere, leaked into the troposphere about one year later and entered the ocean by precipitation. By 1973, about 45% of the total bomb ^{14}C had entered the ocean and penetrated to an average depth of 300 m. In the surface layers, the largest values of both isotopes are located in the Pacific and Atlantic Oceans at about 50°N. The concentrations decrease towards the Southern Hemisphere, but the concentration gradient is not uniform. Near 15°N there is a remarkable reduction in both ^{14}C and ^{3}H concentrations, as shown in Figure 5.15, which is associated with a frontal boundary between the sub-tropical gyre to the north and the equatorial waters to the south. The low concentrations and limited vertical penetration between 15°N and 15°S suggest that water is being upwelled from depths of at least 500 m. Broecker (1981) has shown that this equatorial upwelling flux of $45 \times 10^6 \, m^3 \, s^{-1}$ is comparable with the total production of bottom water in the global ocean. The most remarkable and unexpected finding, shown in Figure 5.15, is the relatively high concentrations of ^{3}H in the North Atlantic Deep Water only 20 years after its surface input, thus indicating a more rapid ventilation of the deeper ocean than had previously been anticipated. In the western Atlantic, the southward bottom boundary current has been identified by its high tritium concentrations.

In addition to radioactive tracers, chlorofluorocarbons (freons) released into the atmosphere over the last 70 years have found their way into the world's oceans. Individual species of freons can be determined reliably, and are providing additional information on the movement of intermediate and deep water throughout the oceans.

6

Observations of Winds and Currents

6.1 Measurement of winds and currents

Both atmosphere and ocean are fluid environments, and the basic techniques of the measurement of winds and currents have close similarities. There are two fundamentally distinct ways of specifying a flow field in the atmosphere and the ocean. One approach, known as the Eulerian description, expresses the flow field at a fixed position in space relative to the Earth's surface. For example, a meteorologist may obtain an Eulerian description of the surface wind field from measurements of the wind direction and speed at an array of weather stations at a particular instant in time. Mathematically, an Eulerian velocity flow can be defined as a vector quantity at a position x, y, z and at a time t:

$$\mathbf{v} = \mathbf{v}(x, y, z, t) \tag{6.1}$$

The co-ordinates in both fluids refer to an eastward direction, x, a northward direction, y, and vertically upwards, z.

The second approach, referred to as a Lagrangian description, measures the position of the chosen element of fluid as a function of time. This method gives information on the time history of a parcel of air or water. The flow is defined by the vector quantity:

$$\mathbf{v} = \mathbf{v}(\mathbf{a}, t) \tag{6.2}$$

where $\mathbf{a}$ is the position vector of the chosen element of fluid with respect to a defined origin. An example of the Lagrangian specification is the trajectory of a balloon moving with the velocity of the air around it.

It is important to note that the two specifications of the flow field are independent, and therefore they give different types of information on atmospheric and oceanic flows. It is not possible to obtain *exact Eulerian* velocities from the trajectories of balloons or floats and, similarly, it is not possible to determine *exact Lagrangian* velocities from Eulerian measurements. However, as will be shown presently, it is possible to make approximate comparisons between the two systems of measurement if sufficient simultaneous measurements of Lagrangian and Eulerian velocities are available.

The choice of measuring system depends on the type of information required. A Lagrangian measuring system using, say, drifting ocean buoys would be appropriate for determining the movement and dispersal of a natural tracer or a pollutant. On the other hand, a weather forecaster prefers the Eulerian system for specifying winds, since he or she is most concerned with prediction over a specific area. As we have seen above, the distribution of fixed synoptic stations essentially gives an Eulerian description of the circulation and it is not surprising that numerical weather-prediction models have been designed on an Eulerian grid system for the solution of the equations of motion.

It is important to note the different conventions for specifying the flow direction in the atmosphere and the ocean. Wind is usually defined by the direction from which it has come, whilst current directions are defined by the direction in which they are going. Hence, a westerly wind is flowing from the west to the east, and this would correspond to an eastward current in the ocean. From here on, the oceanographic convention for both winds and currents is adopted wherever possible.

In both the atmosphere and the ocean, Eulerian velocities are most commonly measured by a rotary device, such as an anemometer or current meter, where the number of rotations of a set of wings or propellers in a given time is measured. The anemometer has to be secured from a tower or mast, and therefore it can only be used for wind measurements near the Earth's surface. Most cup-anemometers are sensitive to wind variations longer than 10 s and have an accuracy of $1 \, \mathrm{m \, s^{-1}}$. Smaller and lighter anemometers can be used for more accurate observations of the wind close to the ground, such as required in micrometeorological experiments. The main difficulty with anemometer measurements is the problem of the alteration of the flow by obstructions, in particular by the tower or mast. For example, a ship may distort the air flow and this will lead to errors in wind measurement which are dependent on the orientation of the ship to the wind direction and the exposure of the anemometer.

The current meter is now the mainstay of modern Eulerian measurements in the ocean. In recent years, considerable improvements in the technical design of current meters have been made, coupled with better mooring and data systems. A commonly used meter is the Savonius meter, which consists of a set of cups which rotate in proportion to the flow through the meter. The direction of flow is obtained from a freely moving direction vane attached in line with the instrument. The Savonius meter can measure current velocities to an accuracy of $2 \, \text{cm} \, \text{s}^{-1}$. Though the original instrument was designed in the 1930s, it has been improved upon by using a magnetic-tape storage system to record velocity and direction at about eight times per rotor revolution. The instrument samples bursts of data every 15 min because of the high-frequency variability of the currents. It automatically vector averages bursts of data to give an accurate measurement. This is important close to the surface where waves produce low-frequency noise. One unfortunate problem with the Savonius rotor is its susceptibility to vertical motion induced by surface waves, and this alone is capable of inducing the rotor to move. The problem does not arise with the vector-measuring current meter which uses two horizontal axis propellers orientated in perpendicular directions. This automatically measures both horizontal components of the flow and it is unaffected by surface wave motion. In all current meters, the orientation of the instrument is measured by a magnetic compass.

A major difficulty in the Eulerian measurement of currents at sea is the provision of a stable platform at a known position. Current measurements from ships are subject to the ship's motion and care must be taken in their interpretation. Most current measurements are now taken from surface and bottom moorings, which are laid by an oceanographic vessel and left in position for up to two years before being recovered. The main features of the moorings are a radio, a light and a float, all necessary to locate the mooring at the end of an experiment. Current meters, and other instruments, are suspended at the required intervals down to the ocean bottom. Moorings usually have between five and ten current meters attached to the wire or the nylon cable. Buoyancy is necessary because of the weight of the instruments and cable, and it is provided by glass spheres encased in plastic containers. Finally, the wire is anchored to the bottom. A feature of some moorings is the acoustic release system which detaches the cable from the bottom anchor when it is activated by a research ship at the end of the experiment. In the early days of development, the recovery rate was about 50%, but now, with the advent of more robust systems, the recovery rate for moorings is about 95%. Some of the problems that have had to be overcome were from fish severing the mooring lines, from corrosion of the mooring wires and from intense surface currents pulling the floats under. Surface moorings have a higher failure rate than sub-surface moorings because of the considerable stresses induced by surface wave motion. Although moorings have proved a cost-effective way of obtaining Eulerian current observations, the movement of the mooring, particularly at or close to the surface, produces errors. This type of motion can be measured by acoustic tracking of the mooring, and it is therefore possible to compensate for it when analysing the current measurements.

Other Eulerian techniques include the determination of the angle of tilt of a wire supporting a known drag. This is useful for measuring current velocities near the surface but, because the drag force increases with depth, it can give misleading determinations of the currents at depth.

Sound velocity measurements can be used to determine Eulerian velocities by two different methods. The first method requires the measurement of the travel time of sound between two hydrophones, A and B, as shown in Figure 6.1. The travel time between A and B is $L/(c+v)$, where v is the average current velocity, c is the speed of sound and L is the distance between A and B. The travel time between B and A is $L/(c-v)$. In an acoustic current meter L is known and thus v may be determined, without knowledge of the velocity of sound c, from the difference between travel times. Of course, the system as shown will only measure the current velocity in the direction of sound waves, and so a second pair of hydrophones is required in the perpendicular direction to measure both components of the horizontal flow. Munk and Wunsch (1983a) have suggested that this technique be used to measure horizontally averaged velocities on ocean-basin scales

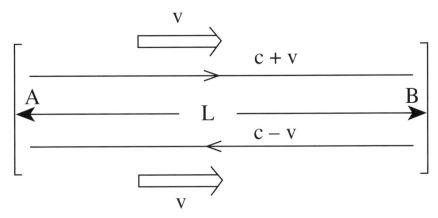

Figure 6.1 Principle of the acoustic current meter. c is the velocity of sound and v the water velocity

by using an array of moored transceivers in the SOFAR channel at a depth of about 1000 m. Such an application would not, strictly speaking, yield truly Eulerian measurements.

The ADCP (Acoustic Doppler Current Profiler) has become a widely-used instrument for the determination of ocean currents. The ADCP can either be mounted on the hull of a ship or on a moored buoy. The sound waves emitted by the ADCP are back-scattered by particles in the water column to the instrument. The relative movement of the particles, and therefore the current which contains the particles, causes a Doppler shift in frequency between the emitted and back-scattered sound.

The velocity of the patch of water, v, is given by:

$$v = \frac{c\Delta F}{F_0} \tag{6.3}$$

where ΔF is the shift in frequency and F_0 is the frequency of the emitted sound.

The current is measured to a depth of about 300 m, with a vertical resolution of 8 m at the surface. In shallow water the ADCP can be referenced to the bottom, and therefore absolute currents are determined reliably. In deep water accurate navigation, provided by the global positioning fixing system (GPS), is required to obtain absolute currents.

In the atmosphere, Eulerian methods include the pressure anemograph. This instrument measures the difference between the dynamic pressure facing the wind and the static pressure. The pressure difference is

$$\frac{1}{2}\rho_A v^2$$

where ρ_A is the density of the air and v is the wind velocity. This technique is very accurate and it is sensitive to rapid fluctuations in the wind. It is able to measure fluctuations of up to 1000 Hz and is particularly useful for turbulence measurements. It can be used, *in situ*, on a mast or tower, or from an aircraft. Horizontal components of the flow can be obtained by using two instruments in perpendicular directions. For very accurate measurements of small-scale turbulence, especially in the boundary layer, a number of techniques are applicable. These include the hot-wire anemometer, which is very sensitive to wind fluctuations, and sonic anemometers. However, it is important to note that, for synoptic data acquisition for weather forecasting, truly Eulerian techniques are limited to surface observations.

Above the Earth's surface, horizontal wind measurement depends on the use of Lagrangian techniques; the most widely used being the tracking of a balloon by radar, theodolite or satellite. During the Second World War, when balloon tracking became important for obtaining measurements of the winds in the upper troposphere for military aircraft operations, small pilot balloons of known positive buoyancy were released from weather stations and tracked by theodolite. The observer could calculate the ascent rate from the known buoyancy of the balloon at the surface and, by accurate timing, could therefore obtain the x, y, z co-ordinates of the balloon at any time. This type of operation could be carried out by single, skilled observers, but it was limited by the height of the cloud base and the assumption of a constant ascent rate. Radar tracking systems permit the range, elevation and azimuth to be obtained automatically, and thus the exact trajectory of the balloon can be traced in all weathers throughout the entire depth of the troposphere and the lower stratosphere. These balloon ascents are not, by definition, exact Lagrangian measurements because the balloon is sampling different parcels of air as it moves upwards through the atmosphere. Furthermore, because it is taking a profile of the atmosphere, the balloon may not be moving exactly with the horizontal velocity of the air at any particular level. However, despite these problems, the balloons provide a valuable source of information on the upper-level winds for weather forecasting purposes. The winds are obtained by taking differences in horizontal positions at, for example, 1 min intervals. A balloon may take about an hour to attain a height of 20 km and, during that time, it may travel a horizontal distance of 50–100 km, depending on the strength of the upper winds. Meteorologists involved in weather prediction are mainly interested in mapping synoptic motions with scales greater than 100 km. By comparison, the horizontal distances travelled by the balloons are small and the measured winds are, therefore, a reasonable approximation to the Eulerian winds at the weather station.

More truly Lagrangian measurements are obtained from constant-level balloons which are designed to float at a specified pressure level in the atmosphere by appropriate adjustment of the buoyancy. These balloons are much larger than their synoptic counterparts and individual balloons have been tracked around the globe using satellite position-fixing systems. In the 1970s, a large number of these balloons were released in the Southern Hemisphere to obtain information on winds over remote regions of the Southern Ocean, where conventional balloon data were unavailable. The balloon trajectories provided valuable information on the circulation of the upper troposphere, but their impact on weather prediction has not been so marked. As mentioned earlier, a major difficulty with a Lagrangian system is that the observer has no control over the position of the balloon, and so the velocity information may not be in the area required by the meteorologist. Another problem was that the formation of ice on the balloon's surface resulted in a high failure rate during the research programme.

A technique for obtaining synoptic-scale wind information over remote areas is to measure the velocity of clouds. High-resolution cloud pictures, in the visible and the infra-red, from geostationary satellites allow the regular monitoring of patches of cloud, on an hourly basis, from which a time-averaged wind speed and direction can be obtained. Furthermore, because many cloud elements can be monitored, a pattern of the large-scale wind circulation can be obtained from a few photographs. The main drawback of the technique is that it is not easy to determine the exact height of the clouds. To obtain the height of the clouds, additional information on the vertical temperature profile and the cloud-top temperature is required.

In the above discussion, it is clear that the main requirement for wind observations has come from the desire to obtain global wind data, in order to understand and predict the weather. The ascents of pilot and radiosonde balloons, coupled with satellite techniques, now provide daily observations of wind at all heights of the troposphere and the stratosphere.

In the ocean, it is only in recent decades that measuring systems have become reliable and frequent enough to obtain time-dependent maps of, admittedly, small parts of the ocean. As noted previously, only recently have current meters been widely deployed in the ocean and, prior to this, most direct information on the large-scale circulation was obtained by Lagrangian measurements as opposed to Eulerian measurements. In the late eighteenth and the nineteenth centuries, with the development of colonial trade routes, information on surface currents was invaluable for minimizing sailing times. Major Rennell, in 1778, was the first to determine the surface currents of the Atlantic Ocean. Providing that the position of the ship, its heading and speed are accurately known, the surface current can be calculated. Later, Maury (1853), the superintendent of the Naval Observatory in Washington, used this technique to produce the first global charts of surface currents. These charts are still the most commonly presented data on surface currents and, with the aid of improved position-fixing systems for shipping using satellite navigation aids, the technique is still valuable for determining large-scale flow patterns, especially where the currents are strong. Another earlier Lagrangian technique, popular for the determination of currents in the nineteenth and early twentieth century, was the drift bottle. Its main disadvantage was that it gave no information about the trajectory between the launch and the pick-up point, and a current determination would always be the minimum velocity of the flow between the two points. Today, drift cards and bottom drifters are still used to determine flow patterns over small distances, particularly in estuarine environments.

The major revolution in modern oceanography has come with the development of remotely tracked ocean buoys. The concept of an ocean buoy system is quite simple, but it is only recently, with improved electronic systems, telemetry and satellite navigation, that they have come into widespread use. A surface drifting buoy consists of a transmitter with batteries and an antenna for radio location by satellite, aircraft or by a shore station. It is desirable to have most of the volume of the buoy below the water to reduce the direct effect of wind stress. Secondly, they usually have a drogue, or parachute, at a depth of 20–30 m to enable the drifter to move with the ocean current rather than with the direct wind-induced current. Many surface drifters also record the surface atmospheric pressure and temperature. Some have a thermistor chain attached which enables the sub-surface temperature profile to be monitored. Surface drifters are less expensive than current moorings and they also give more information on the large-scale flow patterns than an array of current meters. They can be tracked relatively easily for many months, and sometimes for years, and are ultimately limited by their batteries' duration.

It was recognized by Stommel and other oceanographers in the 1950s that a drifting buoy could also yield important information on deep-ocean flows, which were believed to be difficult to measure by conventional techniques. John Swallow, working at the National Institute of Oceanography in the United Kingdom, was the first to design a deep-ocean float which could be tracked by a ship using hydrophones. His success lay in the observation that a body which is less compressible than sea water will gain buoyancy as it sinks. Furthermore, if the excess weight of the body at the surface is small, then at a given depth it will become neutrally buoyant and no further sinking will occur, and thus this float would become the oceanographic equivalent of the constant-level balloon in the atmosphere. The float consists of a 6 m length of aluminium tube, half of which is used for buoyancy, and the other half for housing an acoustic transmitter and batteries. The original float weighed about 10 kg and weights were added to obtain a negative buoyancy of 38 g at the surface, thus allowing the float to become neutrally buoyant at a depth of about 1000 m. Two hydrophones, one at each extremity of the research ship, were used to determine the distance and direction of the float. One of the difficulties is that it is not possible for a ship to track more than a few floats at any one time. Furthermore, the ship is totally involved in locating the floats and this is a very expensive operation. A subsequent development, by Rossby *et al.* in the USA in the early 1970s, involved the location of neutrally buoyant floats in the sound channel, between 1000 and 1500 m, which enabled continuous tracking of large numbers of floats by remote, shore-based hydrophones up to 1000 km away. These SOFAR floats have been located in the western Atlantic for over a decade and a considerable quantity of information on large-scale ocean circulation has now been collected, as evidence by Figure 6.2. It is important to realize that even neutrally buoyant floats only give a picture of the horizontal Lagrangian flow at a specified depth, and that they can give no information on the vertical velocity field. Secondly, a large density of floats is required to give a reliable, statistical picture of the flow field. For example, in the MODE (Mid-Ocean Dynamics Experiment), 20 floats were required to adequately monitor a 300 km diameter region of the ocean west of Bermuda.

Recently, the ALACE (Autonomous Lagrangian Circulation Explorer) float has been developed and deployed in the South Pacific and Southern Ocean. This float is essentially a Swallow float which can drift at a predetermined depth. The float comes to the surface at 25-day intervals, where its data and position can be transmitted to a

Figure 6.2   (a) A plot of 700 m SOFAR float trajectories. Arrows are 100 days apart. Note the high velocity of floats caught in the Gulf Stream

satellite. Currently, 275 floats are deployed at a depth of 1000 m and have a lifetime of about four to five years, depending on the battery supply.

In the meteorological context, both radiosondes and dropsondes (i.e. radiosondes parachuted from aircraft) give a measurement of the vertical variation of horizontal velocity, which is useful for the detection of large local velocities associated with jet streams and internal waves. When used in conjunction with other wind measurements from other radiosonde stations, a three-dimensional picture of the horizontal flow field can be obtained. An oceanographic equivalent of the dropsonde is the velocity profiler, which is simply an acoustic float with temperature and pressure sensors. It is allowed

Figure 6.2 (b) The mean equivalent Eulerian circulation estimated from 700 m SOFAR float data

to sink slowly over an array of hydrophones on the bottom. As with the neutrally buoyant floats, the time delays between the float emission and the hydrophone reception allow the float's position to be determined to an accuracy of ± 1 m. By differencing the position plot, the horizontal velocity can be obtained with an accuracy of ± 2 cm s^{-1}. Figure 6.3 shows some results obtained by Luyten and Swallow using a vertical profiler on the equator in the Indian Ocean. The considerable vertical variation of equatorial currents surprised both experimental oceanographers using conventional current-meter moorings and theoretical oceanographers. It demonstrated, yet again, just how much remains to be observed and understood about the ocean circulation.

All the Lagrangian techniques described above only measure a part of the three-dimensional flow pattern, and therefore represent only quasi-Lagrangian methods. However, it has long been recognized that a chemical or radioactive tracer in a turbulent environment like the ocean will move with the individual parcel of water into which it is inserted, provided that it is in a sufficiently low concentration to leave the physical properties of the water mass unaltered. In Chapter 5 it was shown how naturally occurring chemical and radioactive tracers can be used to infer circulation patterns, but the monitoring and diagnosis of a Lagrangian tracer experiment are quite different to the techniques previously described. Consider the evolution with time of a patch of dye on the ocean surface. The dye spreads in all directions as the result of a large number of small-scale fluctuations in the surface velocity induced, perhaps, by the surface wind or by small eddies in the flow. This is an example of diffusion. Furthermore, the centre of gravity of the dye patch will show a translation. Thus, the dye gives two types of information:

(i)   The area spread of the dye gives information on the mixing by time-dependent currents.
(ii) The movement of the centre of gravity of the dye patch gives information on the mean current.

To monitor a tracer experiment, it is necessary to sample both spatially and with time. Time-lapse photography of the dye patch, from a ship or an aircraft, can give useful information over a small surface area but, below the surface, photography can only be used in the clearest waters. Fluorimeters, towed along beneath the surface of a

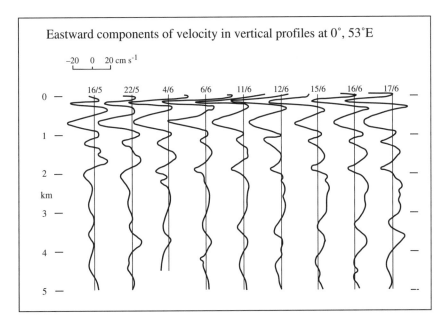

Eastward components of velocity in vertical profiles at 0°, 53°E

Figure 6.3 Horizontal component of velocity obtained from a velocity profiler in the equatorial Indian Ocean

ship, can be used to measure dye concentrations accurately and quickly. A useful, non-toxic dye is rhodamine-β and it can be detected at concentrations of one part in 10^{10} by fluorescence measurements. The most frequent use of dye measurement is in local areas such as coastal seas and estuarine environments where it is used to study the dispersal of pollutants.

Chemical and radioactive tracers require more time-consuming onboard and shore-based techniques to measure concentrations. Care has to be taken to prevent the contamination of samples during collection and transfer to the laboratory, but such tracers do have the advantage of being able to yield high-precision results. Tritium, produced by atmospheric hydrogen-bomb tests in the late 1950s and early 1960s, has been used to trace both horizontal and vertical ocean currents over large areas of the North Atlantic. In the waters around the British Isles, circulation has been determined by the regular monitoring of cassium-137, which is a low-level radioactive waste product from the nuclear reprocessing plant on the Cumbrian coast.

6.2   Scales of motion in the atmosphere and ocean

Figure 6.4 shows four individual time series of wind speed measured by an anemometer. The first series shows the fluctuations in velocity covering a period of one minute, the second shows one minute average values for a period of one hour, the third shows one hour average values for a period of four days and the fourth shows one day average values for a period of one year. The main conclusion that can be drawn from these four time series is that fluctuations in wind velocity occur on a whole range of time scales, from a few seconds to years. Time series, not dissimilar to those of the atmosphere, can also be obtained for the ocean. A meteorologist, or an oceanographer, has to be aware of these fluctuations and their magnitudes. For example, a synoptic meteorologist who wanted to obtain wind information to discern large-scale weather systems would be interested in obtaining 10 min or hourly averaged winds, in order to remove the high-frequency variations which have little direct relevance to large-scale flow patterns. On the other hand, a meteorologist who wanted to know how the smoke from a chimney was going to behave in the immediate vicinity of the discharge would be interested in the high-frequency fluctuations in the wind, because it would be the gusts of wind which would be responsible for the mixing of the smoke and the spread of the plume. A climatologist would probably average winds over a month, a season or even a few years in order to ascertain long period fluctuations and trends.

A useful way to represent these fluctuations is to convert the time series shown in Figure 6.4 into time frequency spectra, as illustrated in Figure 6.5. The simplest approach to this kind of analysis is to consider a time series as the summation of a series of sine and cosine functions having a range of discrete frequencies. The highest frequency is $\pi/\Delta t$, where Δt is the sampling period of data, and the lowest frequency is $2\pi/T$, where T is the period of the observations. Thus, a time series can be represented by:

$$u(t) = u_0 + \sum_{n=1}^{N} a_n \cos nt + b_n \sin nt \tag{6.4}$$

Figure 6.4   Four time series from anemometer; I minute, I hour, 4 days and I year

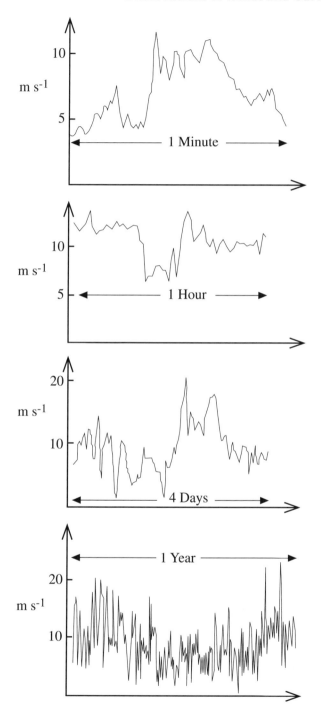

where u_0 is the mean value of the time series. The coefficients a_n and b_n can be determined from the following relationships:

$$a_n = \frac{2}{T} \int_0^T u(t) \cos\left(\frac{2\pi nt}{T}\right) dt$$

$$(6.5)$$

$$b_n = \frac{2}{T} \int_0^T u(t) \sin\left(\frac{2\pi nt}{T}\right) dt$$

where T is the length of record.

A Fourier decomposition reveals two categories of information. First, the modulus of the coefficients, given by $\sqrt{(a_n{}^2 + b_n{}^2)}$, yields a measure of the amplitude of the fluctuations at frequency n, whilst the ratio of the coefficients, $\tan^{-1}(a_n/b_n)$, determines the phase in radians of the fluctuations at frequency N. Though the Fourier method and also other spectral methods using finite data produce a discrete set of frequencies, there is a tendency in the literature to show the spectra as continuous, as in Figure 6.5. As can be seen from this figure, there is no single frequency which dominates the spectrum, although there are certain frequencies where the fluctuations are large. In this atmospheric spectrum there is a peak of fluctuations with a period of 1 min associated with gusts produced by convection processes. There is also

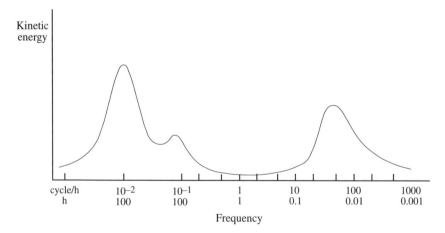

Figure 6.5   Frequency spectrum of the horizontal wind velocity

a tendency for high-energy fluctuations at one day which is caused by diurnal variations in the wind, whilst between the period of one minute and one day the fluctuation energy is smaller. For periods longer than a day there is a dramatic increase and a maximum occurs at a period of a few days. This is due to the large-scale synoptic variability of the atmosphere. Transforming the time series into a frequency spectrum thus makes it possible to perceive different time scales of motion in the atmosphere.

Figure 6.6 shows a frequency spectrum obtained from the analysis of readings from current meters moored in 2000m of water. It is evidently quite different from the atmospheric spectrum, having two identifiable peaks. One peak arises from the semi-diurnal tide (12.4 hours) and the other peak is an inertial oscillation which has a period of 19 hours. The inertial oscillation will be discussed in Chapter 7, but it is essentially a function of latitude and the rotation of the Earth. Its period, T, is given by:

$$T = \frac{\pi}{\Omega \sin \phi} \qquad (6.6)$$

where Ω is the rotation rate of the Earth and ϕ is the latitude. At longer periods the spectrum shows no obvious peak, but does display a large amount of variability. This suggests that a variety of mechanisms may be causing low-frequency time variability in the ocean.

The time variability is only one of several possible dimensions in which to analyse motions. The spatial equivalent is the wavenumber spectrum, the wavenumber being the inverse of wavelength. In this case, unlike the infinite time series, there are quite definite limits to the spectrum which are defined by the finite size of the ocean basins and the Earth's atmosphere. Figure 6.7 demonstrates the wavenumber frequency spectrum of the wind obtained from the analysis of the large-scale atmospheric flow. It can be seen that a large part of the variability is on very long scales, with peak energy in waves which have five to eight cycles around the Northern Hemisphere. The lowest wavenumbers (1–4) are long Rossby waves, and they are associated with the distribution of the continents and oceans, and with topography. Higher wavenumbers, particularly between 8 and 10, are associated with synoptic weather systems, which have space scales of a few thousand kilometres. At higher wavenumbers (i.e.

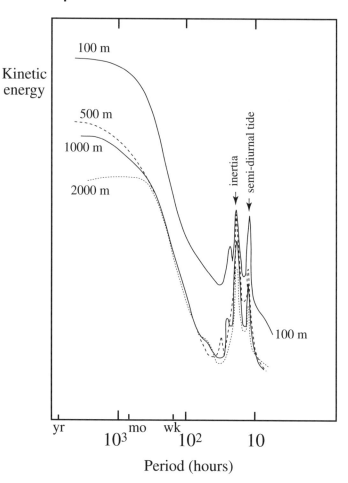

Figure 6.6 Frequency spectrum of kinetic energy of horizontal currents from four depths at Site D (39°N, 70°W) north of the Gulf Stream. Note the two high-frequency peaks, coinciding with the inertial period (19 hours) and the semi-diurnal tidal period (12.4 hours). The kinetic energy associated with the tidal and inertial frequencies is smaller than the low-frequency variability between 7 and 20 days

shorter wavelengths), the energy of the spectrum drops very quickly, indicating that the large-scale systems are, energetically, the dominant flow in the atmosphere. As will be shown later, this is not true of the ocean basins where it is found that the energetically dominant space scales are associated with mesoscale circulations having wavelengths of approximately 100 km.

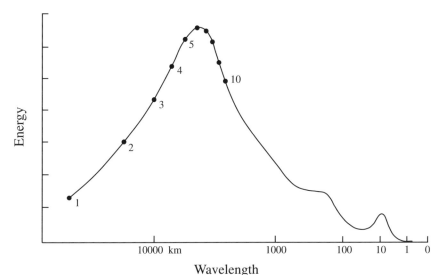

Figure 6.7 Wavelength spectrum of wind fluctuations in temperate latitudes. Numbers indicate circumpolar wavenumbers. Area under segment of curve is proportional to eddy kinetic energy in a corresponding band of the spectrum

6.3   Time-dependent motion

Time-dependent atmospheric motion

It has been shown that time-dependent motions occur on a variety of space and time scales. The larger scales of motion constitute the most energetic and longest time-scale phenomena in the atmosphere. Figure 6.8 shows the trajectory of a constant-level balloon at 15 kPa (height of approximately 15 km) observed over a period of 23 days. If the flow was purely zonal and constant in time, concentric circles would be anticipated with the balloon repeating a similar trajectory on each circuit but, instead, considerable variations in time and space are observed. On the last circuit of the hemisphere there are two large excursions over South America and the eastern Atlantic Ocean which were not apparent on the previous circuit. These deviations from circular motion are caused by Rossby waves. The number of these Rossby loops in the circumpolar flow can vary from one or two to as many as six or seven waves. If they were sinusoidal waves, a spectrum such as the one shown in Figure 6.7 would indicate their wavenumber. For example, a wavenumber 1 flow would be a single wave around the

Figure 6.8 A portion of the trajectories of constant-level balloons at height of 15 kPa (~13 km). Successive positions on the dates indicated are also shown

hemisphere, whilst a wavenumber 4 flow would indicate four wave loops around the hemisphere. The central axis of the waves is often associated with jet streams and, unlike the climatological sub-tropical jet stream with mean winds of $30\,\mathrm{m\,s}^{-1}$, wind speeds of up to $100\,\mathrm{m\,s}^{-1}$ can occur. The observation and forecasting of these jet streams are important priorities for aircraft operators. The low wavenumber patterns tend to move only slowly with respect to the Earth's surface, with the circumpolar flow moving eastward through the pattern. Occasionally, the Rossby wave pattern may become stationary for a few days, or even weeks, and then it is known as a 'blocked flow'. The waves are generated by a dynamic instability of the strong circumpolar flow, which grows on a time scale of a few days. However, the waves may be enhanced by large mountain barriers such as the Andes or the Rocky Mountains, and by the thermal contrast between ocean and continent.

Embedded within these large Rossby wave loops are the familiar, synoptic-scale cyclones and anticyclones which have length scales of the order of 1000–3000 km, and time scales of one to five days. There are typically 20–25 cyclones and anticyclones in the lower part of the atmosphere (i.e. below 5 km) in the Northern Hemisphere at any one time. There tend to be larger numbers of cyclones than anticyclones, but the latter tend to have larger horizontal scales. The extra-tropical, or middle-latitude, cyclone is also associated with stronger wind circulation than the anticyclone. The positions of the cyclones and anticyclones do not tend to be random with respect to the upper air flow. The cyclones form on the eastern side of the trough of the Rossby wave, whilst anticyclones are generally located within, or slightly downstream of, the ridge. Within a stationary or 'blocked' flow, the anticyclones and cyclones will grow and decay repeatedly over the same regions, thus resulting in persistent regions of 'good' and 'bad' weather. The surface pressures of anticyclones are generally in the range of 102–105 kPa, whilst extra-tropical cyclones have a range of 93–101 kPa. Both highest and lowest pressures tend to occur in the winter hemisphere. Vertical motions within synoptic systems tend to be between 1 and 10 cm s^{-1}.

Within the extra-tropical cyclones are frontal systems which range from 50 to 200 km in width. These frontal regions tend to spiral out from the cyclone's centre and they may extend for many thousands of kilometres, and even into another cyclonic system. Frontal regions are boundaries separating air masses and are, therefore, associated with large temperature gradients, especially in winter. Temperature changes are usually between 5 and 10°C, but occasionally, over continental regions, temperature changes of up to 20°C are observed. These frontal regions are often dynamically active, and they are associated with upward motion and continuous precipitation. The strongest upward motion in an extra-tropical cyclone is usually associated with the frontal systems and it may reach values of 0.5 m s^{-1} in intense fronts.

The next important scale in atmospheric motions is the mesoscale, which includes a variety of phenomena ranging from shower clouds, thunderstorms and sea-breeze circulations to larger systems such as squall lines and rain bands. The former group of phenomena has scales of 1–10 km and time scales of the order of one hour, whilst the latter may reach 100 km in size and persist for up to one day. The squall line is often associated with a line of thunderstorms and strong, gusty winds, and it is common in tropical latitudes. Rain bands are regions of heavy precipitation which exist in, and propagate through, larger mid-latitude frontal systems. The tropical cyclone, typhoon or hurricane is a large mesoscale system, of the order of 100 km in diameter, associated with intense winds of up to $50 \, \mathrm{m \, s^{-1}}$ and heavy precipitation. Vertical velocities in mesoscale systems may vary from $1 \, \mathrm{m \, s^{-1}}$ in a tropical cyclone or a sea-breeze circulation to more than $10 \, \mathrm{m \, s^{-1}}$ in intense thunderstorms.

Smaller-scale systems, less than 1 km in size, tend to last no longer than a few minutes. The most ferocious small-scale phenomenon is the tornado which, although less than $\frac{1}{2}$ km in diameter, may attain circulation velocities of between 50 and $100 \, \mathrm{m \, s^{-1}}$. Vertical velocities tend to be of a similar order of magnitude. Waterspouts are small-scale funnel clouds which form over the sea. They are usually less intense but more numerous than tornadoes. Both tornadoes and waterspouts tend to have a cyclonic rotation which is derived from the circulation of the parent thunderstorm cloud. Small-scale whirlwinds, including the dust devil, may attain wind speeds of $10 \, \mathrm{m \, s^{-1}}$ in a column of a few tens of metres wide and about a hundred metres high, persisting for a couple of minutes. They are formed in the surface layer when the prevailing wind is light and the surface is being intensely heated. The temperature gradient close to the surface becomes very large, or superadiabatic, and the surface layer becomes unstable. The strong upward acceleration results in the spectacular growth of a vortex. Other small-scale phenomena include the circulation within a cumulus cloud, and the local eddying motion around buildings and obstructions. Beyond the small-scale phenomena are the microscale motions associated with the final dissipation of kinetic energy into thermal energy by viscosity.

Time-dependent oceanic motion

Ever since surface current maps (see Figure 6.13) were first produced in the nineteenth century, the variability of ocean currents has also been recognized. Indeed, on surface charts, information is included on the steadiness of the currents, in addition to their mean speed and direction. As detailed in Section 6.2, both current meter and float measurements have indicated the presence of variability not only in the surface layers of the ocean, but also in the deepest abyssal layers. A small part of the variability is accounted for by predictable frequency variations such as tides and by inertial motion. Reference to Figure 6.6 will show that the majority of variations occur on time scales greater than a few days, with a peak energy between 50 and 200 days. Beyond this time scale there is an indication of a reduction in energy, but suitable time series (i.e. time series sufficiently long to distinguish these low-frequency currents) are rare. However, there are seasonal fluctuations in currents such as the Gulf Stream and, more spectacularly, the Somali current, which reverses with the monsoon, and thus variations at the annual period may be anticipated. On even longer time scales, low-frequency wind forcing, from interannual and climatic variations, should be expected to produce very long-period fluctuations in the ocean but, as yet, such fluctuations have not been clearly distinguished.

The first recognition of the variability of the deep ocean came with the Aries experiment west of Bermuda in 1959 when 72 neutrally buoyant floats were individually tracked for 4–10 days over an experimental period of 14 months. It had been expected, from indirect measurements of hydrographic properties, that the floats would travel northwards with a steady velocity of approximately $1\,\mathrm{cm\,s^{-1}}$. However, the floats drifted away with velocities of $10\,\mathrm{cm\,s^{-1}}$ in all directions, much to everyone's surprise. From the apparently random pattern of the buoy tracks, it was possible to deduce that energetic eddies with length scales of 50–100 km were present. Subsequent to this early experiment, many neutrally buoyant and SOFAR floats have been tracked, and a more general picture of the mean deep-ocean flow has emerged. Figure 6.9 shows the trajectory of a single SOFAR float over a period of one year in the North Atlantic during the MODE experiment. It can be seen that the float is caught in Eddy A for 27 days and Eddy B for 28 days. The diameter of Eddy A is about 60 km, and therefore the mean speed around the eddy is about $10\,\mathrm{cm\,s^{-1}}$. This float also shows a tendency to drift towards the south-west with a gentler speed of about $1\,\mathrm{cm\,s^{-1}}$ over the year. When many more floats were tracked and their trajectories plotted, though the picture was very complicated, the westward drift of the floats was very discernible. Current meter data

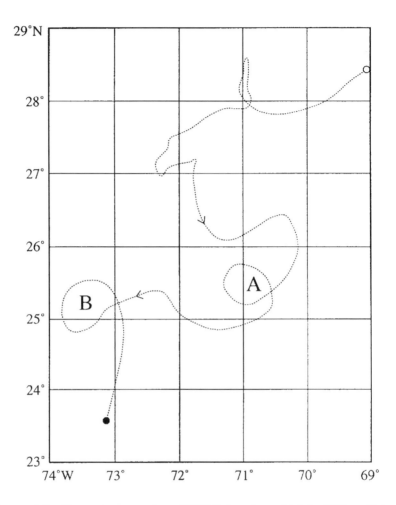

Figure 6.9   The trajectory of one SOFAR float from 11 July 1973 to 30 June 1974. It is caught in each eddy for about one month. Note the general westward drift of the float of about 1 cm s^{-1}

from moorings in the same region show current speeds of between 20 cm s^{-1} close to the surface to 10 cm s^{-1} near the bottom. The variability of the current has periods ranging from three to about six weeks, which is typical of mesoscale eddies. These measurements, taken in a relatively small region of the Atlantic Ocean, have shown that not only is there variability in the ocean, but also that the current variability is about one order of magnitude larger than the mean velocities of the flow. This is due to mesoscale eddies, and thus the kinetic energy per unit volume of the eddy is probably about 100 times larger than that of the mean circulation. Similar studies in other regions of the North Atlantic Ocean and in other ocean basins are revealing energetic, mesoscale eddies but, as yet, a detailed picture of the eddy field for all the ocean basins has not emerged. Long-term float measurements, which have been averaged over 100 days to remove the eddy field, have shown details of the large-scale circulation. Figure 6.2a clearly shows the trajectories of floats in the anticyclonic flow of the large sub-tropical gyre (i.e. clockwise, in the Northern Hemisphere). A few floats have been caught up in the intense jet-like flow of the Gulf Stream. The velocity of one such float has a mean velocity of $25 \ \mathrm{cm \, s}^{-1}$ in a 100-day interval.

Other important scales of variability exist in the ocean. On the larger scale, seasonal and long-period wind variations can give rise to time variations in currents, such as those observed in the Indian Ocean. Only recently have long-period current measurements been obtained, but they have already shown that remarkable oscillations in supposedly steady currents do exist. In 1974, current moorings on the equator showed significant 15-day oscillations in the flow of the equatorial undercurrent. The meanders are apparently the result of a hydrodynamic instability of the undercurrent and it has also been found that it varies on an interannual time scale, in response to changes in wind circulation. In 1982–83, the equatorial undercurrent in the eastern Pacific Ocean virtually disappeared for a period of two months as a result of a reversal in the surface wind from westwards to eastwards. Meanders in surface equatorial currents, about 1000 km long, have been observed by infra-red satellites measuring sea surface temperatures. Furthermore, large eddies have recently been discovered in the Somali current during the south-west monsoon. These eddies are rather larger than the mesoscale eddies described previously. They are about 500 km in diameter and tend to be slower moving.

On smaller scales, less than 100 km, surface frontal systems are observed which are energetic for their scale. Fronts form around coastal upwelling features, and form the boundary between the cold upwelled water and the warmer surrounding water. They are also found along the edge of the continental shelf break at the boundary between oceanic water and the shelf water. Convergent and divergent surface flows will also tend to result in frontal discontinuities, and may be associated with surface wind patterns as, for example, the Antarctic divergence and the sub-tropical convergence zones. Such flows may also be associated with a developing eddy such as a Gulf Stream ring, where a frontal discontinuity develops about the edge of the ring separating the warm Sargasso Sea water from colder, fresher shelf water.

Small regions of deep convection have been discovered in the Mediterranean and Labrador Seas and the Antarctic Ocean, and result from intense surface cooling. The convection occurs on horizontal scales of 10 km and may extend down to the ocean bottom, thus providing a source of deep, cold water. Measurements with a vertical current meter have shown vertical velocities of 10^{-2} m s^{-1} in these convection regions. Such velocities are of the order of 100 000 times larger than the mean vertical velocity of the general ocean circulation, as inferred from tracers. However, these deep convective events are relatively infrequent and very localized.

6.4   Climate and seasonal circulation

Climate and seasonal time-averaged circulation of the atmosphere

Care has to be taken in the interpretation of the time-averaged pictures of circulation systems as complex and variable as the ocean and the atmosphere. For the purposes of clarity, the atmospheric mean circulation will be defined as the climatic monthly average taken over a period of 5–10 years. For surface observations, a 30-year average is usually taken as the climatic average, but there is no *a priori* reason to suppose that this will be more representative of the circulation than a 10-year average. Figure 6.10 shows the longitudinally averaged winds as a function of height and latitude for the months of January and July, obtained from daily observations of the atmosphere from weather stations, ships, aircraft and balloons. The predominant feature in both hemispheres is a region of high wind speed, reaching 30 m s^{-1}, in the upper troposphere between 30 and 40° latitude. These climatological, eastward-flowing jet streams are the most energetic part of the tropospheric circulation. It is noted that the Northern Hemisphere sub-tropical jet stream moves poleward in the summer and weakens by about 50%. A similar poleward movement occurs in the Southern Hemisphere, but the

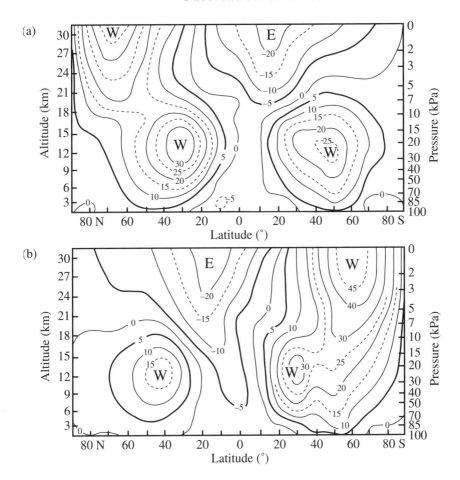

Figure 6.10 Mean zonal wind for (a) December–February, (b) June–August. Units: m s^{-1}. Positive (negative) values denote westerly (easterly) flow

seasonal variation in intensity is smaller. The magnitude of the jet stream is related to the horizontal meridional temperature gradient and this is discussed in Chapter 7. The mean jet flow is dominated by eastward winds throughout most of the upper troposphere, but westward winds do occur in the tropical troposphere and near the surface in polar regions. It can be appreciated that it is necessary to have both eastward and westward winds at the surface to avoid a net wind torque on the Earth which would result in a change in the Earth's rotation.

In the stratosphere there is a very important seasonal variation in winds. There is a general tendency for the eastward jets to weaken immediately above the troposphere, but in winter a strong jet develops at high latitudes with magnitudes of up to $70\,\mathrm{m\,s^{-1}}$. This is known as the polar night jet and it is caused by the large meridional temperature gradient between the polar and the sunlit stratosphere (Figure 6.11). As discussed in Chapter 3, the temperature of the stratosphere is controlled by the absorption of solar radiation by ozone. During the polar night, in the absence of solar heating, there is a rapid decline in the temperature of the polar stratosphere. In the polar summer, the meridional temperature gradient is reversed because maximum radiation occurs at the poles and the stratospheric wind circulation in the summer hemisphere is everywhere westwards. This situation is depicted in Figure 6.10.

A different picture of the mean circulation is obtained from Figure 6.12, which shows the vertical and meridional components of the flow. A large cell, known as the Hadley cell, is seen to cover the tropical region and two much weaker cells, known as the Ferrel cell and polar cell, respectively, are also visible. In the Hadley cell, air is lifted up over the equatorial zone, mostly in the ITCZ, where large amounts of latent heat are released by heavy precipitation. This air moves polewards to the sub-tropics (30° latitude), where it descends. During its descent the air warms by adiabatic compression and its relative humidity decreases. The circuit is completed by a return flow in the lowest $2\,\mathrm{km}$ of the atmosphere towards the convergence zone. This return flow is very steady and it is associated with the north-east and south-east trade-wind systems. Some of the descending air in the sub-tropics travels polewards and ascends in the middle latitudes. This ascent occurs in the region of the sub-polar low-pressure systems. A weak return flow to low latitudes completes the circuit. One feature of the mean meridional circulation is that it is much weaker than the corresponding zonal circulation, having mean velocities of less than $3\,\mathrm{m\,s^{-1}}$, and the maximum velocity occurs in the low-level trade winds. The vertical velocities are generally less than $1\,\mathrm{cm\,s^{-1}}$. The mean meridional circulation is, therefore, a much weaker feature than the zonal circulation, which is related to the dominance of the equatorial–polar temperature gradient in controlling the flow.

The horizontal climatological surface flow is shown in Figure 6.12. It can be seen that the sub-tropical high pressures are the 'sources' of

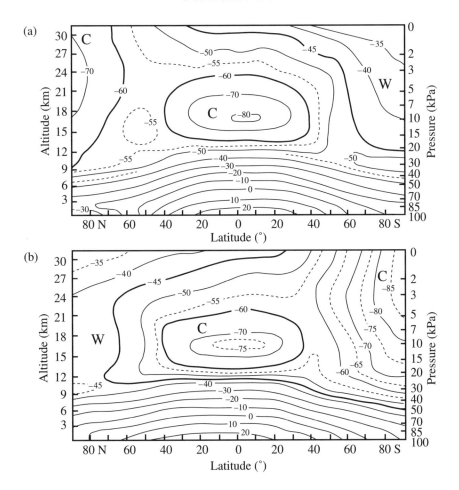

Figure 6.11 Mean temperature for (a) December–February, (b) June–August. Units: °C. W = warm; C = cold

the circulation. These are regions where descending air from the upper branch of the Hadley cell enters the lower troposphere and they are largely cloud-free, with only low-level broken cloud usually observed. The trade winds can be seen flowing towards the equatorial zone where they meet, from both hemispheres, at the ITCZ. These convergence zones are generally located in the Northern Hemisphere in the Pacific and Atlantic Oceans along the thermal equator (i.e. the region of maximum sea surface temperature).

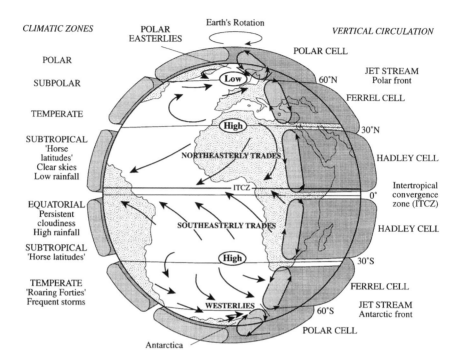

Figure 6.12   Schematic representation of features of the general circulation of the atmosphere

In the Indian Ocean and the western Pacific Ocean, seasonal monsoonal variations in the wind circulation dominate the flow. The large mass flow from the southern Indian Ocean high to the Asian continent is evident in summer, from about June to September. The curving of the winds from a south-east flow in the Southern Hemisphere to a south-west flow in the Northern Hemisphere is related to the rotation of the Earth, as discussed in Chapter 7. In winter, the low-pressure centre is located over the Indonesian maritime continent, and convergence of air from both the Indian and Pacific Oceans takes place. The regions of convergence indicate areas of upward motion and heavy precipitation. Weaker monsoonal circulations occur over Africa and South America.

Climate and seasonal circulation of the ocean

The major data source for the climatic determination of surface ocean currents has been ships' drift measurements. A chart, such as the one shown in Figure 6.13, probably represents well over 100 years of data obtained from merchant vessels. The main features of the chart include:

(i)   The locations of the most intense currents in the equatorial zones and along the east coasts of Japan (the Kuroshio current), North America and South Africa (Agulhas current), where peak velocities reach over $1\,\mathrm{m\,s}^{-1}$.

(ii)   The tendency of mid-ocean currents to be weaker. Surface velocities of $10{-}30\,\mathrm{cm\,s}^{-1}$ are recorded.

(iii) The east–west nature of the currents in the equatorial Pacific and Atlantic.

(iv)   The domination of an eastward current around the globe in the Southern Ocean.

(v)   The alignment of the current boundaries, or frontal regions, in an east–west orientation.

(vi)   The presence of large anticyclonic flows at approximately 30° latitude in all ocean basins except for the northern Indian Ocean.

(vii) The response of the surface currents in the Indian Ocean to the monsoons.

(viii) The small cyclonic circulations occurring in the north-western Atlantic and in the North Pacific near the Aleutian Islands.

It is unlikely that more information on surface currents will drastically alter the above description of the major features. However, below the surface it is not possible to obtain a climatic picture from direct measurements, except over small regions. It is only in the last 25 years that current meter moorings have been deployed in significant numbers and of sufficient duration to obtain long-term mean currents. With the deployment of deep-ocean drifters, a much improved description of the mean circulation is being obtained, but the present picture of the global circulation at depth is still incomplete. However, hydrographic data on temperature, salinity and pressure can be used to obtain an indirect estimate of the mean current. The details of this method will be discussed in Chapter 7 but, in its simplest form, it involves the calculation of the horizontal pressure field from the density of sea water and the use of the hydrostatic equation. Except on the equator,

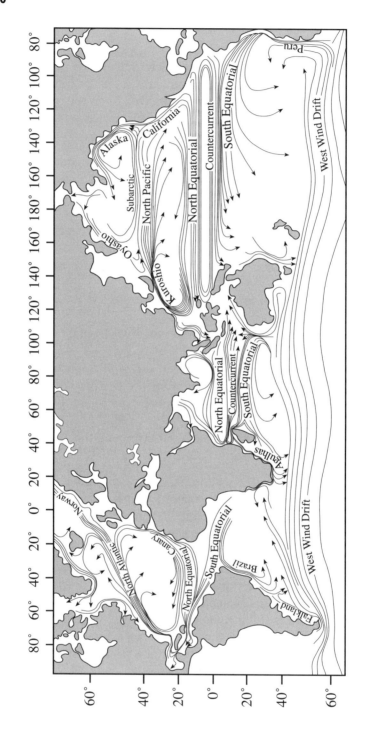

Figure 6.13 (a) Schematic chart of the climatological average of the surface currents of the ocean. (b) Monsoonal surface currents: February. (c) Monsoonal surface currents: August

(b)

(c)

the horizontal flow tends to follow lines of constant pressure, and hence the circulation can be calculated. This flow is known as a geostrophic current. However, only relative horizontal motion can be obtained because it is not possible to calculate the absolute horizontal pressure field. Generally, a reasonable estimate of the upper ocean flow can be obtained by referencing it to a level of, say, 1 km below the surface, where the flow and the horizontal pressure gradients are generally weaker. This method can utilize the substantial hydrographic database for the global ocean to obtain an estimate of the geographical distribution of ocean currents.

Another important feature of the mean flow can be seen in the North Atlantic where a broad poleward flow, called the North Atlantic current, enters the Norwegian Sea and the Arctic Basin. This returns from the Arctic Basin into the North Atlantic via the intense east Greenland boundary current. However, not all of the poleward flow is compensated by return flows between the surface and 1500 m, and this implies that the sinking of water must occur for conservation of mass. This sinking occurs in parts of the Norwegian Sea, the Greenland Sea and the Labrador Sea. However, because of the bottom topography and the episodic nature of this sinking, the abyssal flow is not a continuous, steady, southward current. It tends to be most intense close to the western edge of the western Atlantic basin.

The observations of vertical motions in the ocean are, at best, scanty. Figure 6.14 shows a schematic vertical section across the Atlantic Ocean for the uppermost 800 m of the ocean, obtained from a variety of indirect methods. In the surface layers the eastward and westward wind-driven currents are evident. The surface currents produce regions of convergence at approximately 30º latitude in both hemispheres in the centres of the sub-tropical gyres and this results in downward motion in the upper 200–400 m which, in turn, is partially responsible for the dip in the thermocline at these latitudes. A large fraction of this water is subsequently transported upwards and equatorwards along the main thermocline to the equatorial upwelling region, where the thermocline is often less than 100 m below the surface. In the Northern Hemisphere two small, closed cells are found between the equator and 10ºN, where downwelling and a dip in the thermocline are produced. The dipping of the thermocline induces an eastward geostrophic flow between the generally westward equatorial current, and is known as the equatorial counter current.

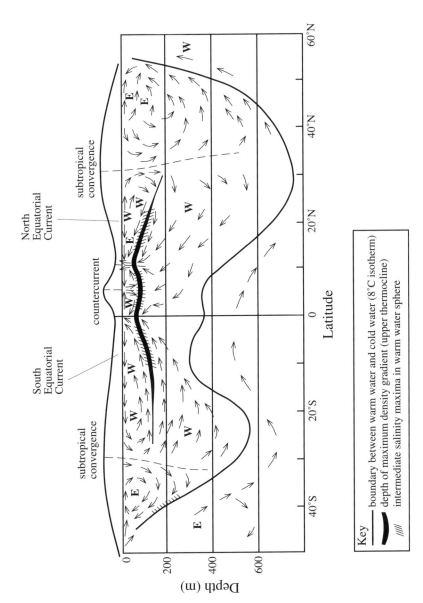

Figure 6.14 Schematic meridional cross-section through the upper layers of the Atlantic Ocean

The vertical flows in the middle and tropical latitudes of the open ocean are induced by the surface currents. In polar latitudes, the sinking of water and movement along isopycnal surfaces induce vertical circulations. The equatorial upwelling zone may produce vertical velocities of up to $10^{-4}\,\mathrm{m\,s^{-1}}$, whilst sub-tropical downwelling regions have velocities of less than $10^{-6}\,\mathrm{m\,s^{-1}}$. In the deep ocean, vertical motions are associated with velocities of approximately $10^{-7}\,\mathrm{m\,s^{-1}}$, although local values of $10^{-3}\,\mathrm{m\,s^{-1}}$ have been recorded in very small sinking regions less than 10 km in diameter. In these regions the use of tracers has proved to be invaluable in deducing the vertical flow. Figure 5.15 shows the tritium distribution in the North Atlantic about 10 years after its surface input ceased. This figure clearly shows the weakness of the vertical circulation in the Atlantic Ocean. Deep sinking is evident at high latitudes in the North Atlantic but, in general terms, all of the remaining tritium is confined to the waters above the main thermocline. The upward bowing of the distribution is evident in the equatorial zone where upwelling occurs.

From this brief résumé of the mean ocean circulation, it is clear that more remains to be discovered than is known at present. However, by using different indirect techniques of determining current flow, it has been shown that the climatological flow is much weaker in the deep ocean than at the surface. The surface flows are determined, to a substantial degree, by the surface wind distribution, and seasonal changes of the wind circulation can result in changes in the surface circulation, as seen in the Indian Ocean. The deep flows are not directly determined by the surface winds and are more dependent on the vertical sinking of polar waters (the thermohaline circulation), the horizontal density gradient and the geometry of the ocean basin.

7

The Influence of the Earth's Rotation on Fluid Motion

7.1 An introduction to the Earth's rotation

The Earth is a rapidly rotating planet. A point on the Earth's surface at the equator has a velocity of $463 \, \mathrm{m \, s^{-1}}$ relative to an absolute frame of reference such as the position of the stars. By comparison, our nearest planetary neighbours Venus and Mars have absolute velocities on their equators of 1.8 and $239 \, \mathrm{m \, s^{-1}}$, respectively. The velocities of the atmosphere and ocean relative to the rotating Earth are small, and therefore a parcel of air or water will only travel a small distance on the globe compared with the movement of the Earth. Under these circumstances, it is to be expected that the rotation of the Earth will have a profound influence on the circulation of both atmosphere and ocean.

First, consider the effect of rotation on a stationary particle on the Earth's surface. The particle will be subjected to the acceleration due to gravity, g, and a centrifugal acceleration in the direction perpendicular to the axis of rotation, as illustrated in Figure 7.1. At the point P, the outward centrifugal acceleration is $\Omega^2 r_1$, where $r_1 = r_e \cos \phi$. At the equator this acceleration is $0.034 \, \mathrm{m \, s^{-2}}$, compared with an average acceleration due to gravity of $9.81 \, \mathrm{m \, s^{-2}}$. The resultant vector, known as the 'effective gravity', is directed at an angle of 0.1° to the true direction of gravity. In both meteorology and oceanography, it is assumed that the direction of gravity is that of the resultant gravity vector, and therefore the centrifugal acceleration is implicitly taken into account. The local vertical is also assumed to be aligned with the resultant gravity vector.

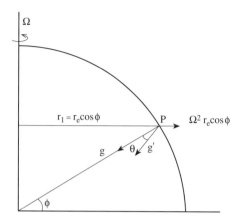

Figure 7.1   Centrifugal acceleration on a particle (P) stationary with respect to the Earth. **g'** (apparent gravity vector) is the resultant of the centrifugal acceleration and gravity, **g**. The radius of the Earth is r_e and the latitude is ϕ

In Chapter 1 it was shown that a geopotential surface is a surface of constant gravitational energy and, by definition, the component of gravity is perpendicular to this surface. Due to the variation of the local component of g with position, it is necessary to define a level surface in terms of a geopotential surface rather than a geometric surface. The geopotential, Φ, is the work per unit mass in moving a parcel of fluid a vertical distance, z, against gravity and is given by:

$$\Phi = \int_0^z g \, dz'$$ (7.1)

where z is the geometric height relative to mean sea level. It is noted that the geopotential has units of energy. The acceleration due to gravity varies horizontally from the pole to the equator by about 0.5% and in the vertical by 0.23% per 10 km from sea level.

In both oceanography and meteorology, a common unit for the geopotential is the dynamic meter. In the atmosphere it is defined as:

$$Z = \Phi/10$$ (7.2)

and in the ocean as:

$$D = -\Phi/10$$ (7.3)

The negative sign appears because it is measured downwards from the surface of the ocean.

The geopotential, or dynamic height, is related to the pressure, p, by the hydrostatic equation:

$$\alpha \, dp' = -g \, dz' = -d\Phi \qquad (7.4)$$

For the ocean, the dynamic height at pressure p with reference to sea level in a stationary ocean is:

$$\int_p^0 \alpha \, dp' = -\int_h^0 g \, dz' = -\Phi \qquad (7.5)$$

$$\therefore \quad D = \frac{1}{10} \int_p^0 \alpha \, dp' \qquad (7.6)$$

It is assumed that the pressure at sea level is zero in the ocean. The geopotential depth, D, can therefore be calculated from numerical integration of the profile of the specific volume, α, obtained from *in situ* measurements of temperature and salinity.

The influence of rotation on a moving particle on the Earth's surface will now be considered and the following example demonstrates how rotation will affect the motion of a projectile, as perceived by an observer on a rotating frame. Consider a turntable rotating with an angular velocity Ω, as shown in Figure 7.2. If a person stands at the centre and throws a ball, the ball will take a curved path with respect to an observer on the turntable. It will appear to this observer that the ball has rotated in a direction opposite to that of the turntable. To a stationary observer looking down on the turntable, the ball will appear to have been unaffected by the rotation of the turntable and it will be perceived to move in a straight line. The deflection of the ball is, therefore, the result of the transformation from a non-rotating to a rotating frame of reference. This deflection force is known as the Coriolis force.

In Figure 7.2 the ball is launched in the direction of OA_2, with an initial velocity v. After a small increment of time, Δt, the ball has reached the edge of the turntable but, because of the rotation of the turntable, point A_1 has rotated to A_2. According to an observer at A_1, the ball has been

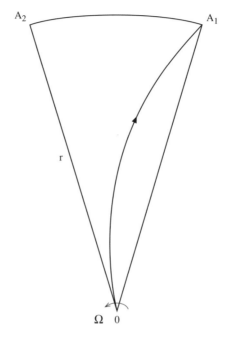

Figure 7.2 Coriolis effect: deflection of a moving particle on a turntable

deflected a distance A_1A_2, perpendicular to the initial direction of the ball. If the radius of the turntable is r, then:

$$A_1A_2 = \Omega \Delta tr \qquad (7.7)$$

and since $r = v\,\Delta t$, then by substitution in equation 7.7

$$A_1A_2 = v\Omega\,\Delta t^2 \qquad (7.8)$$

A particle under a field of constant acceleration, a, will be displaced by $\frac{1}{2}a\,\Delta t^2$ and therefore, by inspection of equation 7.8:

$$a = 2\Omega v \qquad (7.9)$$

First, it is noted that the Coriolis acceleration, a, is proportional to the velocity in a rotating frame of reference, and therefore a stationary body on the Earth will not be subjected to a Coriolis acceleration. Secondly, the Coriolis acceleration is a vector which acts perpendicular to both the direction of motion of the projectile and to the direction of the axis of rotation, which is vertical in this case. The vector acceleration, **a**, is given by the vector product:

$$a = 2\mathbf{\Omega} \times v \qquad (7.10)$$

On a rotating planet the axis of rotation is only aligned in the direction of the vertical axis at the poles, whilst elsewhere it is at an angle of $90 - \phi$, where ϕ is the latitude, as shown in Figure 7.1. Therefore, to obtain the Coriolis acceleration on a locally horizontally moving particle, it is necessary to take the component of rotation in the direction of the local vertical axis. From Figure 7.1, this component is $\Omega \sin \phi$ and thus, for horizontal motion, the Coriolis acceleration is dependent on the latitude. At the equator the local vertical is perpendicular to the axis of rotation and therefore, for horizontal motion, the Coriolis acceleration is absent. It will be noted that, in general, vertical motion will be subjected to a Coriolis acceleration as a result of the component in the local horizontal direction. However, because of the small vertical accelerations observed in both atmosphere and ocean, and the dominance of gravitational and vertical pressure accelerations, this component of the Coriolis force is usually negligible.

For a moving particle, relative to a rotating Earth, an additional centrifugal component will arise. However, because of the relatively low velocity of the atmosphere and ocean, this centrifugal component is small and it is often neglected. For a horizontal motion of $10\,\mathrm{m\,s^{-1}}$ at 45° latitude, the Coriolis acceleration is $1.03 \times 10^{-3}\,\mathrm{m\,s^{-2}}$, which compares with a centrifugal acceleration of $2.21 \times 10^{-5}\,\mathrm{m\,s^{-2}}$. It can therefore be seen that the centrifugal force is only about 2% of the Coriolis force. However, both forces are small compared with the acceleration due to gravity.

7.2 Inertial motion

A practical demonstration of the existence of the Earth's rotation can be made by observing the rotation of a Foucault pendulum. Figure 7.3 shows a pendulum oscillating along a line AA' at the North Pole. Assume that there is an observer a small distance from the pole on the Greenwich Meridian. The pendulum will be unaffected by the Earth's position, but the observer will be rotated with the Earth. To the observer, the plane of the pendulum's oscillation will apparently rotate in the opposite direction to the rotation of the Earth. After six hours the observer will have rotated by 90° and will be aligned with the

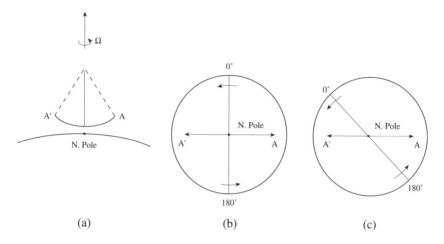

Figure 7.3 Foucault pendulum: (a) oscillating at the North Pole, (b) and (c) apparent rotation of the pendulum relative to a rotating Earth

pendulum, and after 12 hours the pendulum will be observed to oscillate in the original direction. This period of the rotation of the pendulum is known as the pendulum day or the inertial period. At the poles the period is 12 hours but, because it is determined by the component of rotation along the local vertical axis, it will increase as the latitude decreases. At a latitude ϕ, the period of rotation will be 12/sin ϕ hours. At the equator, an east–west oscillating pendulum will be aligned in the same direction as the rotation of the Earth, and thus there will be no apparent rotation relative to the Earth. However, if a pendulum oscillated in a north–south plane across the equator, it will tend to rotate in an anticlockwise direction in the Northern Hemisphere and in a clockwise direction in the Southern Hemisphere. It will thus produce a figure of eight pattern.

It will now be shown that parcels of fluid on a rotating Earth will exhibit very similar behaviour to that of the pendulum. In a non-rotating system, from Newton's second law, a body will continue in a state of uniform motion unless acted upon by a force. For inertial motion in an absolute frame of reference, the rate of change of velocity of the fluid parcel is given by:

$$\frac{D_a}{Dt}(\mathbf{v}) = 0 \tag{7.11}$$

where the subscript a refers to an absolute frame of reference. In Section 7.1 it was shown that there is an additional Coriolis acceleration in a rotating frame of reference, where the equivalent relation to equation 7.11 may be written as:

$$\frac{D_r}{Dt}(\mathbf{v}) + 2\mathbf{\Omega} \times \mathbf{v} = 0 \qquad (7.12)$$

where $(D_r/Dt)(\mathbf{v})$ is the acceleration of a fluid element as measured on the rotating Earth.

Equation 7.12 may be written for two horizontal components of motion:

$$\frac{Du}{Dt} - fv = 0 \qquad (7.13)$$

$$\frac{Dv}{Dt} + fu = 0 \qquad (7.14)$$

where u is the velocity in the eastward direction, v is the velocity in the northward direction and $f = 2\Omega \sin \phi$. The subscript r, denoting a rotating frame of reference, has been dropped and, from now on, it will be assumed that all motion is relative to the rotating Earth. The main point to be noted here is that a fluid element, or any other body, cannot continue in a state of uniform motion on a rotating Earth, even if all frictional forces and pressure gradient forces could be eliminated.

Equations 7.13 and 7.14 will now be used to deduce the trajectory of a fluid element initially moving northwards with a velocity v in the Northern Hemisphere where $f > 0$. From equation 7.13 it will be seen that the northward motion will induce an acceleration towards the east, because the Coriolis acceleration in the Northern Hemisphere always acts to the right of the direction of motion. The parcel's trajectory will curve towards the east, because of the eastward acceleration, and it will obtain a positive u component of velocity. From equation 7.14 the Coriolis force will induce a southwards, or negative, acceleration which will decrease the original northward component. At some stage the parcel will have zero northward velocity and it will then have only an eastward component. After this stage it will continue to accelerate southwards and it will acquire a negative v component of velocity which, from equation 7.13, will induce a westward acceleration in its

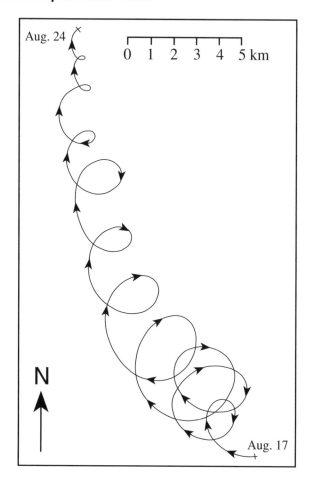

Figure 7.4 Progressive vector diagram illustrating an inertia current superimposed on a northwards motion, observed in the Baltic Sea between 17 and 24 August 1933

turn. Thus, the parcel will follow a circular trajectory and, providing that the change of latitude of the parcel is very small, it will return to its original position. This circular trajectory is known as an inertia circle.

Eliminating v from equations 7.13 and 7.14 yields:

$$\frac{D^2 u}{Dt^2} + f^2 u = 0 \qquad (7.15)$$

This is the equation for simple harmonic motion with frequency f, the Coriolis parameter. The period of oscillation is:

$$T_i = \frac{2\pi}{f} \text{ or } \frac{\pi}{\Omega \sin \phi} \tag{7.16}$$

It can be shown that the period of the motion is $12/\sin \phi$ hours, as for the Foucault pendulum.

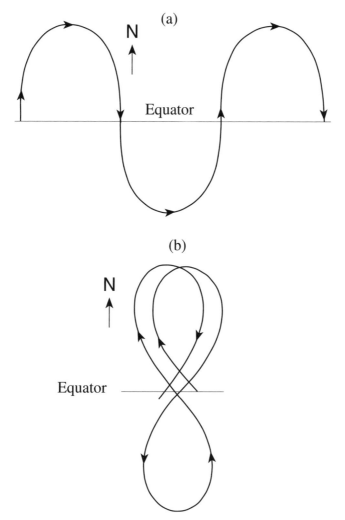

Figure 7.5   Theoretical examples of inertial oscillations about the equator. (a) Initial northward impulse. (b) Initial north-west impulse

Inertial oscillations are frequently observed in the ocean and they can always be identified by spectral analysis of a current meter time series, as shown in Chapter 6. Figure 7.4 shows an example of an inertial oscillation recorded by a current meter in the Baltic Sea, identified by plotting a progressive vector diagram. The theoretical period of 14 hours 8 min was confirmed by these observations. In the atmosphere, inertial observations are only usually obtained at times when pressure gradients are weak and there is a stable vertical stratification which reduces the effects of surface friction. These conditions tend to occur in anticyclonic flow at night. The oscillations will increase the local wind speed when moving in the direction of the large-scale flow, but they will decrease it when moving in the opposite direction. Such oscillations can produce local nocturnal jets in the atmospheric boundary layer which have velocities of up to twice the large-scale geostrophic flow. Figure 7.5 shows theoretical examples of inertial flow in the atmosphere where the motion is assumed to extend over a large enough range of latitude to produce a significant change in the Coriolis parameter, f. It is noted that a variety of patterns across the equator occur, depending on the initial speed and direction of the fluid element. None of the trajectories form closed loops because of the latitudinal variation of f.

7.3   Pressure gradients and geostrophic motion

The principal force acting in both atmosphere and ocean is the pressure-gradient force. It has already been shown that, in the vertical direction, the dominant balance of forces, known as the hydrostatic balance, is between the upward pressure-gradient force and the downward gravitational force. In the horizontal plane the principal balance of forces is between the Coriolis force and the horizontal pressure-gradient force. This balance is known as geostrophic equilibrium.

The horizontal equation of motion, assuming that there is no friction, is now given by:

$$\frac{Du}{Dt} - fv = -\frac{1}{\rho}\frac{\partial p}{\partial x} \qquad (7.17)$$

$$\frac{Dv}{Dt} + fu = -\frac{1}{\rho}\frac{\partial p}{\partial y} \qquad (7.18)$$

The right-hand side terms are the horizontal components of the pressure-gradient acceleration and they can be derived in a similar way to that shown in Section 2.4.

For non-accelerating flow on a rotating Earth, Du/Dt and Dv/Dt are both zero, and so:

$$-fv_g = -\frac{1}{\rho}\frac{\partial p}{\partial x} \qquad (7.19)$$

$$+fu_g = -\frac{1}{\rho}\frac{\partial p}{\partial y} \qquad (7.20)$$

where u_g and v_g are the eastward and northward components of the geostrophic velocity, respectively.

Consider a situation in the Northern Hemisphere where the atmospheric pressure at sea level decreases southwards, as shown in Figure 7.6a. The pressure-gradient acceleration will also be directed southwards since $(1/\rho)(\partial p/\partial y)$ is positive and, to produce a balance of forces, an equal and opposite directed Coriolis force is required acting northwards. Since the Coriolis force is perpendicular to the direction of flow (i.e. to the right in the Northern Hemisphere), a westward geostrophic flow will result. In the Southern Hemisphere the Coriolis force acts to the left of the motion, therefore a southward pressure gradient will result in an eastward geostrophic flow (Figure 7.6b). In general, clockwise flow occurs around Northern Hemisphere high-pressure and Southern Hemisphere low-pressure systems, whilst anticlockwise flow occurs around Northern Hemisphere low-pressure and Southern Hemisphere high-pressure systems. Normally, the term anticyclonic flow refers to the flow around a high-pressure centre and cyclonic flow to low-pressure centres. A simple rule for geostrophic motion is that the Coriolis force is always directed, following the motion, 90° to the right in the Northern Hemisphere and 90° to the left in the Southern Hemisphere. As the Coriolis force is directed against the pressure gradient, the high-pressure centre must always be in the same direction as the Coriolis force.

(a)　　　　　　　　　　　　　　　　　　(b)

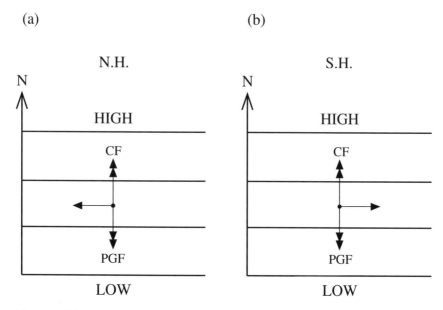

Figure 7.6 Balance of forces in geostrophic flow: (a) in the Northern Hemisphere, (b) in the Southern Hemisphere. CF is the Coriolis force and PGF is the horizontal pressure-gradient force

It must be remembered that geostrophic flow in the atmosphere and in the ocean is only an approximation to the actual horizontal flow. For example, rapid changes in the pressure gradient may occur which would induce inertial accelerations and non-geostrophic flow. In addition, frictional drag of the flow over the Earth's surface will induce deviations from geostrophic flow. On the equator there is no Coriolis force, and therefore geostrophic flow is impossible. In the atmosphere as a whole, geostrophic flow is a good approximation, being accurate to about 10%, to the observed flow above the atmospheric boundary layer (i.e. approximately 1 km above the Earth's surface) in large- and synoptic-scale systems whose linear dimensions exceed 1000 km. In smaller-scale systems, such as thunderstorms and sea breezes, there are large departures of the winds from geostrophic motion. In the upper troposphere in intense jet streams, where velocities can reach $100\,\mathrm{m\,s^{-1}}$, the flow is not geostrophic. In the ocean, in contrast to the atmosphere, the geostrophic approximation can be applied to smaller-scale systems such as mesoscale eddies with a length scale greater than 50 km, as well as to large-scale flows. However, significant departures from geostrophic motion do occur in the surface mixed layer of the ocean due to local wind-driven currents, and also within 100 m of the bottom, particularly where there are large variations in bottom topography.

Some examples of geostrophic motion in the atmosphere and in the ocean will now be considered. Equations 7.19 and 7.20 can be used directly in conjunction with observations of pressure on a horizontal surface to obtain the geostrophic wind in the atmosphere. Accurate measurements, to within about 10 Pa, can be made of the surface pressure using mercury barometers. Horizontal pressure gradients in middle latitudes are of the order of 1–5 kPa over a distance of 1000 km. Geostrophic winds can therefore be calculated to an accuracy of at least 1%. In practice, pressure observations have to be corrected to mean sea-level pressure (101.3 kPa) using the hydrostatic equation. This correction may introduce further sources of error because a temperature has to be assumed for the fictitious air mass between the ground height and sea level. Above the Earth's surface, pressure measurements are obtained from aneroid barometers inserted into a radiosonde package. In these expendable packages, the accuracy is only about 100 Pa. Of course, the height of the balloon at a particular pressure has also to be determined before the geostrophic wind equation can be used. From the simultaneous temperature measurements made by the radiosonde package, the hydrostatic equation (equation 7.4) can be numerically integrated from the surface at a known pressure p_0 to the observed pressure p. The geopotential height of each pressure measurement is thus obtained. Another method of deducing the horizontal pressure gradient above the ocean surface, illustrated in Figure 7.7, involves making simultaneous measurements of pressure and height from an aircraft. Radar altimeters allow geometric height measurement to an accuracy of 0.1 m, but this technique is ultimately limited by deviations of the sea surface from the mean sea level as a result of horizontal variations in the geoid and of ocean currents.

In the ocean there are difficulties in obtaining absolute geostrophic velocities. The first, and most difficult, problem lies in the measurement of the horizontal pressure gradient. In recent decades, bottom-pressure recorders have been deployed in the Atlantic and Indian Oceans. Their principal objective has been to obtain short-period records of pressure variations for tidal analysis. However, with longer-term measurements, it has been found that there is a drift in the instrument calibration

Figure 7.7   Horizontal pressure gradients measured by an aircraft over the ocean before (lower curve) and after (upper curve) correction for altitude variations. The resulting pressure difference, Δp, can be used in geostrophic wind calculations

which is of the same order as the longer-period pressure changes. On the ocean bottom, one is trying to detect pressure variations between 100 and 4000 Pa, compared with an absolute bottom pressure of about 4×10^7 Pa, over periods of months and possibly years. This implies that the bottom-pressure recorder has to become more sophisticated, and presumably expensive, if this experimental problem is to be successfully overcome. In the meantime, oceanographers have developed an indirect method of obtaining geostrophic currents known as the dynamic method. Unfortunately, this method has one important deficiency in that it can only be used to determine relative geostrophic currents. The method is based on the fact that, although it is not practical to obtain an absolute measurement of the pressure gradient at one level in the ocean, it is possible to determine the pressure gradient at one level *relative* to a second level by use of the hydrostatic equation. The vertical pressure difference between the two levels is dependent on the density, or specific volume, of the sea water which can be determined from temperature and salinity measurements. It has been shown in Section 7.1 that the geopotential difference between two pressure surfaces can be obtained from the integration of the specific volume, α, between the two surfaces. In order to use this information to calculate geostrophic currents, the pressure-gradient acceleration along a constant geopotential surface has to be transformed into the geopotential slope of a constant pressure, or isobaric, surface.

The horizontal pressure-gradient acceleration is given by $(1/\rho)\delta p/\delta x$. For hydrostatic equilibrium:

$$\delta p = \rho\,\delta\Phi \tag{7.21}$$

where $g\,\delta z = \delta\Phi$, and therefore:

$$\left(\frac{1}{\rho}\right)\frac{\delta p}{\delta x} = \frac{\delta\Phi}{\delta x} \tag{7.22}$$

In the limit as $\delta x \to 0$:

$$\frac{1}{\rho}\left(\frac{\partial p}{\partial x}\right)_z = \left(\frac{\partial\Phi}{\partial x}\right)_p \tag{7.23}$$

where $(\partial\Phi/\partial x)_p$ is the slope of the constant pressure surface, p.

The geostrophic current along a pressure surface p_1 is therefore given by:

$$fv_1 = \left(\frac{\partial\Phi_1}{\partial x}\right)_{p_1} \tag{7.24}$$

However, it is not possible to determine $\partial\Phi_1/\partial x$ because there is no reference geopotential surface. Although the sea level of a stationary ocean would be a geopotential surface, currents and tidal variations produce significant departures from the geoid. The geostrophic current at a second pressure level, $(p_2 > p_1)$, is

$$fv_2 = \left(\frac{\partial\Phi_2}{\partial x}\right)_{p_2} \tag{7.25}$$

subtracting equations 7.25 and 7.26

$$v_2 - v_1 = \frac{1}{f}\frac{\partial}{\partial x}(\Phi_2 - \Phi_1) \tag{7.26}$$

From equation 7.4 the difference in geopotential, $\Phi_2 - \Phi_1$, between the two pressure surfaces is:

$$\Phi_2 - \Phi_1 = -\int_{p_1}^{p_2} \alpha\,dp \tag{7.27}$$

or alternatively:

$$\Phi_2 - \Phi_1 = -10\,\Delta D \qquad\qquad (7.28)$$

where ΔD is the dynamic height between the two pressure surfaces measured in dynamic meters. A dynamic meter is equivalent to $10\,\text{kPa}$ and is approximately $1.02\,\text{m}$ depth. Thus:

$$v_1 - v_2 = \frac{10}{f}\frac{\partial}{\partial x}(\Delta D) \qquad\qquad (7.29)$$

A similar equation can be derived for the u component of flow.

Considerable effort over the last century has been devoted to the independent determination of v_1 or v_2. This can be done by:

(i) Measurement of the current v_2 at a given depth.
(ii) Determination of a depth of no motion, i.e. $v_2 = 0$, from, for example, a tracer experiment.
(iii) Measurement of the total transport through a confined passage using, say, an underwater cable (see Section 6.1). This requires a vertical profile of the relative geostrophic velocity from the surface to the bottom, and the profile is obtained by repeated applications of equation 7.29. The geostrophic velocity at each depth can then be adjusted by a constant velocity to give the observed transport over the total depth.

Method (i) can only be used where long-period measurements are available. Method (ii) has been used widely, but it often gives contradictory results. In some regions, such as the Antarctic Circumpolar Current, there is an absence of a level of no motion. Method (iii) has been used to determine the geostrophic flow in the Gulf Stream across the Florida Straits, but it is not of general use in the open ocean.

A recent development is the satellite radar altimeter which can measure the elevation of the sea surface to an accuracy of at least 2–3 cm. Figure 7.8 shows the elevation of the sea surface over a north-west to south-east section of the Gulf Stream, measured by SEASAT. The large slope at 37°N is the Gulf Stream itself and the dip in elevation at 34°N is a Gulf Stream ring, or a cold mesoscale eddy. The sea surface slope across the Gulf Stream is $1.30\,\text{m}$ in $125\,\text{km}$ or 1.04×10^{-5}, which is equivalent to a surface geostrophic velocity of $1.16\,\text{m\,s}^{-1}$. This technique may, at last, be able to determine the absolute geopotential slope and provide a reference velocity for geostrophic calculations.

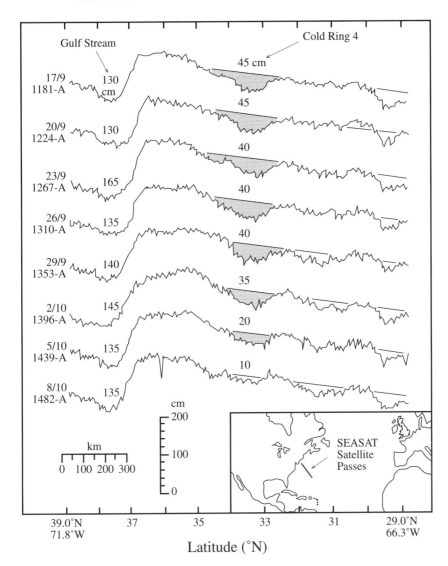

Figure 7.8   Altimeter residuals for eight collinear passes obtained over a three-week period by SEASAT in 1978 in the Gulf Stream region

However, the method will require accuracies of 1 cm, rather than the present 2–3 cm, to measure the sea surface slopes in mid-ocean where geostrophic velocities are an order of magnitude less than those of the Gulf Stream.

Another recent technique, used in the MODE experiment, is to use a combination of float information, current meter readings and dynamic heights to obtain an objective analysis of the absolute geostrophic flow. Figure 7.9 shows a sequence of mesoscale eddies over a period of 60 days in a 500×500 km box in the western North Atlantic Ocean. The flow direction can be determined in exactly the same way as for atmospheric geostrophic flow, i.e. anticyclonic flow around the high-pressure areas and cyclonic flow around the low-pressure areas.

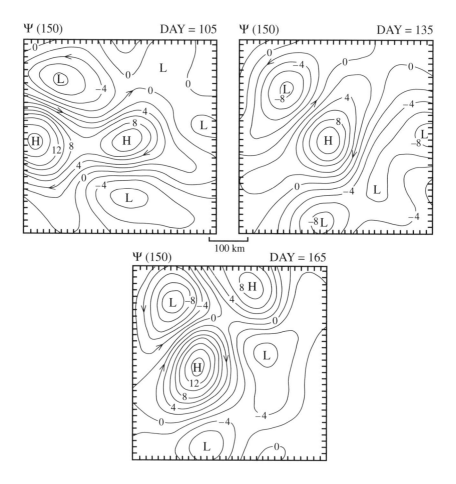

Figure 7.9   The horizontal geostrophic flow at three times, each separated by 30 days, at a depth of 150 m. Tick marks on the side indicate the interpolated grid interval (15 km) and the map centre is at 69°40′W, 28°N

7.4   Vorticity and circulation

A useful property of fluid flows in both ocean and atmosphere is the spin, or vorticity, of a fluid element. Vorticity is defined as the spin of an infinitesimal element of fluid about its own axis, as illustrated in Figure 7.10. It is a vector quantity and directed along the axis of rotation perpendicular to the surface of the fluid element. Positive vorticity is defined as anticlockwise rotation. In both the ocean and the atmosphere, the large-scale flow is predominantly horizontal because vertical velocities are, in general, much smaller than horizontal velocities. It is, therefore, the vertical component of vorticity which arises from the horizontal motion, which is most useful for aiding an understanding of the larger-scale motions.

The vertical component of vorticity, ξ, is defined as

$$\frac{\partial v}{\partial x} - \frac{\partial u}{\partial y}$$

(a)

(b)

(c)

Figure 7.10 (a) Vertical component of vorticity, ξ, in a horizontal flow. (b) Vorticity in a sheared flow. (c) Vorticity in a curved flow

Figure 7.10 shows that vorticity can arise from two distinct flows:

(i) Circular flow.
(ii) Shear flow.

In an absolute frame of reference, in the absence of vorticity sources and sinks, a tube of vorticity is conserved as it moves through the fluid:

$$\frac{D_a \eta}{Dt} = 0 \qquad (7.30)$$

In the rotational frame of reference, the vertical component of absolute vorticity, η, is the sum of the relative vorticity (i.e. the vorticity seen by an observer on the Earth's surface) and the planetary vorticity due to the spin of the Earth. For horizontal fluid motion the planetary vorticity is the component of vorticity in the direction of the local vertical, which is the Coriolis parameter, f, where $f = 2\Omega \sin \phi$. Note that the vorticity of solid-body rotation is 2Ω, where Ω is the angular velocity. In the rotational frame of reference, equation 7.30 becomes

$$\frac{D_r}{Dt}(\xi + f) = 0 \qquad (7.31)$$

where ξ is the relative vorticity.

Consider a non-rotational tube of fluid on the equator. The tube has neither relative nor planetary vorticity. However, if the tube is transferred polewards, it must gain planetary vorticity appropriate to its new latitude. According to equation 7.30, its absolute vorticity, as perceived by a non-rotating observer, must remain zero, and so the tube of fluid must have gained a component of relative vorticity of the same magnitude as its planetary vorticity but opposite in sign. In the Northern Hemisphere the Coriolis parameter, f, is positive, and therefore the acquired relative vorticity is negative and the rotation of the tube of fluid is clockwise or anticyclonic. Hence, changes in relative vorticity will be induced by meridional flow due to the conservation of absolute vorticity.

The following two examples will illustrate the application of this conservation principle to flows in the ocean and in the atmosphere. First, consider a simple poleward movement of fluid which initially has zero relative vorticity at latitude ϕ_0. As shown in Figure 7.11, the fluid will gain relative anticyclonic rotation as it progresses polewards in the

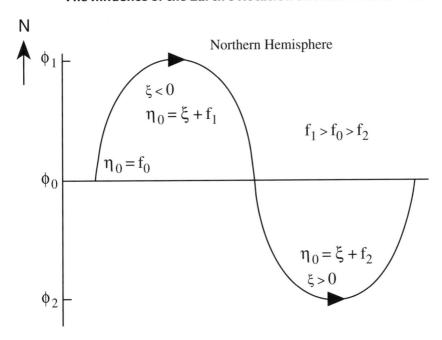

Figure 7.11 Conservation of absolute vorticity in a Rossby wave (see text)

Northern Hemisphere. The anticyclonic vorticity will induce an anticyclonic curvature in the parcel's trajectory. This will cause the parcel to turn first eastwards, then southwards. As the parcel moves equatorwards it will tend to lose its anticyclonic relative vorticity and, on reaching its original latitude, it will again have zero relative vorticity and its direction will be equatorwards with no curvature. At lower latitudes it will gain cyclonic relative vorticity, in order to maintain the absolute vorticity at its original latitude, and this will, in turn, induce a cyclonic curvature in the trajectory which will eventually rotate the parcel's motion back to a poleward direction. It can therefore be seen that the interchange of relative and planetary vorticity, required to conserve the original absolute vorticity, will produce a horizontal oscillation about the original latitude. These oscillations are known as Rossby waves, or planetary waves, and they can be readily identified in the trajectories of constant-level balloons in the atmosphere (Figure 6.8).

The Lagrangian derivative in equation 7.31 can be expanded into the Eulerian derivative form thus:

$$\frac{\partial \xi}{\partial t} + u\frac{\partial \xi}{\partial x} + v\frac{\partial \xi}{\partial y} + \beta v = 0 \qquad (7.32)$$

where $\partial \xi/\partial t$ is the local change in relative vorticity, and the second and third terms represent the transport of relative vorticity by the horizontal flow. $\beta = \partial f/\partial y$ and thus βv is the change in relative vorticity caused by meridional velocity, v.

Consider a simple distribution of meridional velocity given by:

$$v = v_0 \sin (\omega t - kx) \qquad (7.33)$$

where ω is the frequency and k is the wavenumber in the x direction. The wavelength is $2\pi/k$. The relative vorticity, ξ, is given by:

$$\xi = \frac{\partial v}{\partial x} = -kv_0 \cos (\omega t - kx) \qquad (7.34)$$

Substituting into equation 7.32:

$$+k\omega v_0 \sin (\omega t - kx) + 0 + 0 + \beta v_0 \sin (\omega t - kx) = 0$$

Thus:

$$\omega k + \beta = 0$$

or

$$\omega = -\beta/k \qquad (7.35)$$

The phase velocity is

$$c = \frac{\omega}{k} = -\frac{\beta}{k^2} \qquad (7.36)$$

The phase velocity of the Rossby wave is always westwards and it increases in magnitude as the wavenumber of the wave decreases. Table 7.1 shows the phase velocity for different wavelengths at 45° latitude. In the middle-latitude atmosphere, the mean tropospheric flow is eastwards, and therefore the Rossby waves will move either eastwards or westwards relative to the Earth, depending on the relative magnitudes of the tropospheric flow velocity and the Rossby wave velocity. If the tropospheric flow is greater than the wave speed, the Rossby wave will move eastwards relative to the Earth

Table 7.1 Phase speeds and periods of Rossby waves for typical wavelengths, derived from equation 7.36

Wavelength	Period (days)	Phase speed (m s^{-1})
100	284	−0.004
1000	28	−0.41
5000	5.64	−10.25
10 000	2.8	−41.00
20 000	1.4	−164.00

and, if it is less, then the Rossby wave will travel westwards. Very long planetary waves will, therefore, tend to travel westwards and short waves will tend to move eastwards. Should the eastward tropospheric flow match the westward wave speed, the wave will become stationary. The most likely stationary wavelengths lie between 5000 and 10 000 km because their wave speeds correspond to the typical mean wind speeds in the troposphere, as shown in Figure 6.10. In the ocean, the large-scale Rossby waves (i.e. those with wavelengths greater than 1000 km) will generally move westwards because any opposing flow in the ocean will be slower than that in the atmosphere.

A second example of the application of the vorticity equation (equation 7.31) yields an explanation of the intense boundary currents which are located at the extreme western edges of the ocean basins and which include the Gulf Stream, the Kuroshio and the Agulhas. First consider the ocean circulation induced by the surface wind distribution. In sub-tropical latitudes, overall all of the major ocean basins, there are semi-permanent, anticyclonic atmospheric flows. This anticyclonic wind stress will also drive an anticyclonic circulation in the upper 500 m of the ocean, by virtue of the vorticity input at the sea surface. The expected response of the ocean would be a mirror image of the overlying wind system and the maximum dynamic height would be in the centre of the ocean basin, with a broad poleward flow in the western half of the basin and an equatorward flow in the eastern half of the basin .Clearly, these expectations do not correspond to reality but, if the relative vorticity changes caused by the initial 'mirror image' flow are considered, it will be seen that the poleward flow always gains anticyclonic relative vorticity, and the equatorward flow will always lose anticyclonic relative vorticity. The equatorward flow in the eastern half of the basin will therefore tend to partially negate the anticyclonic vorticity input by the wind system. In the western half of the basin the poleward flow will always enhance the anticyclonic vorticity input of the wind. In time, the anticyclonic vorticity will become more concentrated in the western part of the basin and the gyre circulation will move from its central position towards the west. This will enhance the poleward flow in the western portion of the basin and weaken the equatorward flow in the eastern part of the basin, as illustrated in Figure 7.12. Finally, an intense poleward boundary current will form, with very large anticyclonic vorticity, against the western boundary of the basin. The question may be asked – what stops the poleward current from becoming more and more intense, as anticyclonic vorticity continues to be pumped into the ocean? In fact, as more vorticity is added, the intense western boundary current becomes dynamically unstable and energetic mesoscale eddies form. These eddies eventually move out of the intense system, and dissipate the energy and vorticity of the western boundary current. A number of major, national research programmes are devoted to furthering the understanding of the role of mesoscale eddies in these western boundary currents.

It can also be shown that an equatorward-flowing western boundary current will form for a cyclonic wind circulation. In this case, the

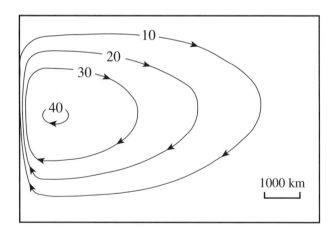

Figure 7.12 The westward intensification of wind-driven currents in a rectangular ocean of constant depth, induced by an anticyclonic surface wind stress. Streamlines show volume transport in $10^6 \, \text{m}^3 \, \text{s}^{-1}$

poleward flow on the eastern part of the wind-driven gyre will gain anticyclonic vorticity and it will therefore tend to oppose the cyclonic vorticity input by the wind. At the same time, equatorward flow on the western flank of the gyre will gain vorticity of the same sign as that of the wind. The narrow, southward Labrador current, on the western boundary of the North Atlantic sub-polar gyre, is an important example of an equatorward-flowing western boundary current. This intense, cold current is responsible for the rapid movement of icebergs from the Greenland glaciers into one of the major shipping lanes between North America and northern Europe.

The vorticity equation (equation 7.31) is only applicable to homogeneous, horizontal flows of constant depth. However, in both the atmosphere and the ocean, vertical motion can have a significant influence on the pattern of vorticity, and therefore on the circulation. First, consider a simple example in which a cylinder of fluid is rotating with an angular velocity Ω. If the cylinder is now stretched along the vertical axis, the diameter of the cylinder will decrease. This will reduce the moment of inertia of the cylinder and, to conserve angular momentum, the cylinder must increase its angular velocity. An analogous situation occurs when a vertical tube of vorticity, having an absolute vorticity η, is stretched along its vertical axis by the vertical velocity gradient $(\partial w/\partial z)$, as shown in Figure 7.13. This can be written as:

$$\frac{D\eta}{Dt} = \eta \frac{\partial w}{\partial z} \qquad (7.37)$$

It is noted that stretching, $\partial w/\partial z > 0$, will always increase the original vorticity, whilst compression, represented by $\partial w/\partial z < 0$, will reduce it. In the absence of vorticity, i.e. $\eta = 0$, vertical motion can neither generate nor destroy vorticity. On a rotating Earth it has been shown that $\eta = \xi + f$, and hence:

$$\frac{D}{Dt}(\xi + f) = (\xi + f)\frac{\partial w}{\partial z} \qquad (7.38)$$

An interesting aspect of this vorticity equation is that relative vorticity can be produced by vertical stretching or compression even in the absence of an original relative vorticity because of the vorticity of the rotating Earth.

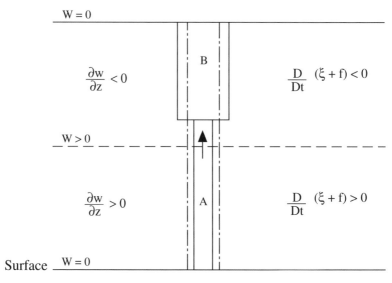

Figure 7.13 Absolute vorticity changes caused by upward motion in the troposphere. It is assumed that the tropopause acts as a lid to vertical motion

Consider first an example from the atmosphere. Figure 7.13 shows a simple upward motion field, induced by surface warming. If the upper troposphere acts as a 'lid' to upward motion (a simplistic assumption), then it can be seen that stretching of the lower troposphere and compression of the upper troposphere would occur. If it is assumed that there is no original relative vorticity (i.e. $\xi = 0$), then there will be an increase of positive absolute vorticity in the lower layer and an increase of negative absolute vorticity in the upper layer of the troposphere. For small synoptic-scale motions, the change in f with latitude can be ignored, and therefore an increase in anticyclonic vorticity in the lower layer is expected. Similarly, an increase in anticyclonic vorticity in the upper troposphere may also be anticipated. This type of behaviour is observed over continental regions in summer when surface heating induces a surface cyclonic circulation. Furthermore, the upward motion implies that air near the surface will converge into the circulation and, if this air is very moist, latent heat will be released on ascent. This will increase the buoyancy of the air and it will intensify the vertical and horizontal motions. Such behaviour occurs in tropical cyclones over warm seas where very moist air converges into the cyclone and aids its development. Although other factors are responsible for initializing the tropical cyclone, this effect becomes dominant in its subsequent growth. It is noted that a sinking motion in the troposphere will have the opposite effect, and will produce an anticyclonic flow in the lower layer and a cyclonic flow in the upper layer. It can therefore be seen that vertical motion can generate both cyclonic and anticyclonic motions in the atmosphere.

On the larger, planetary scale, the change of the Coriolis parameter with latitude becomes an important term in the vorticity balance. In a steady state, where there is no growth of circulation and the relative vorticity is smaller than the Coriolis parameter, then equation 7.38 can be reduced to:

$$\beta v = f \frac{\partial w}{\partial z} \qquad (7.39)$$

This equation says that the relative vorticity produced by vertical stretching is balanced by the relative vorticity lost by meridional motion. Hence, for steady poleward motion, vertical stretching of the lower troposphere will occur and, for equatorward motion, vertical compression will take place. Therefore, large-scale poleward movement of warm air will be associated with rising motion and the equatorward motion of cold air will be associated with descending motion. This simple picture goes some way towards explaining the patterns of vertical motion seen in large-scale Rossby waves, usually called baroclinic Rossby waves. Regions of upward motion occur on the upwind side of a pressure ridge and they are associated with the development of surface cyclonic circulations. Surface anticyclones tend to develop on the downwind side of the ridge where downward motion occurs.

Equation 7.39 has been applied in the ocean by Stommel and Arons (1960) to deduce the deep-water flow. It has been remarked earlier that regions of deep sinking motion in the ocean are very limited in horizontal extent and therefore, to balance this local sinking, upward motion is required over the remainder of the ocean. The observation of a permanent thermocline is evidence of the vertical motion of cold water opposing the downward mixing of warmer water. Hence, in the abyssal layers of the ocean, $\partial w/\partial z$ is positive and, from equation 7.39, poleward motion is expected. This result ran contrary to earlier ideas of deep-water circulation which had suggested that cold, dense water, formed by sinking in high latitudes, would spread uniformly equatorwards over the width of the ocean basin. Stommel and Arons reasoned that, if cold water moved polewards in the interior of the ocean as predicted by equation 7.39, then it must move equatorwards in a narrow deep western boundary current.

In such a narrow, intense flow, equation 7.39 would no longer be valid because the effects of variations in bottom topography on the vorticity balance would become important. Stommel (1958) produced a theoretical model of the abyssal circulation of the world's ocean, driven by sources of deep water in the North Atlantic Ocean and the Weddell Sea, and this is shown in Figure 7.14. Subsequent evidence from all of the ocean basins has shown that these deep western boundary currents do exist. Their velocities are, however, modest when compared with surface western boundary currents. They are generally less than $10 \, \text{cm s}^{-1}$ and the widths of the currents are generally between about 100 and 200 km. There are regions where the circulation shown in this figure is misleading. In the South Atlantic Ocean, the North Atlantic Deep Water moves southwards along the western boundary, but this southward flow overrides the colder and more dense Antarctic Bottom Water which flows northwards along the

Figure 7.14 A schematic diagram of the abyssal circulation of the world oceans, driven by cold deep-water sources in the North and South Atlantic. If the two sources, $\oplus$, in the North and South Atlantic are taken as $20 \times 10^6 \, \text{m}^3 \text{s}^{-1}$ each, then transport between streamlines is of the order of $5 \times 10^6 \, \text{m}^3 \text{s}^{-1}$

same western boundary from the Weddell Sea. The bottom topography also influences the direction of the flow to a marked extent and the Antarctic Bottom Water, near the equator, moves from the western boundary eastwards to the mid-Atlantic ridge. There is also evidence of potentially cold water in the eastern basin of the Atlantic Ocean, near the equator, which may be derived from Antarctic Bottom Water flowing through a gap in the mid-Atlantic ridge. Recent studies, using floats and deep-current moorings below the Gulf Stream, have demonstrated that a deep southward boundary current exists in the region, but its flow is complicated by a series of permanent eddy circulations. These eddies may be driven by the highly unstable flow in the Gulf Stream, but this has not yet been proven.

7.5   The atmosphere and ocean boundary layers

In the previous sections the flows of both the ocean and the atmosphere have been discussed in terms of the geostrophic equation. However, close to the interface between the atmosphere and the surface of the globe, the flow becomes turbulent. The predominantly horizontal flow becomes broken up into vertical eddies which carry momentum down from the free atmosphere towards the surface. These eddies produce a tangential stress on the land or the ocean when close to the surface. This stress can be seen on the sea surface when a gust of wind produces a ripple pattern, known as cats paws. The stress on the surface, τ, is given by:

$$\tau = \rho_A C_D (u^2 + v^2)^{\frac{1}{2}} \mathbf{v} \qquad (7.40)$$

where $\mathbf{v}$ is the mean horizontal wind velocity measured 10 m above the surface. ρ_A is the density of air and C_D is a drag coefficient.

For neutral stability over a sea surface, $C_D = 1.5 \times 10^{-3}$ but, if there is strong vertical convection, the drag coefficient must be increased. A wind of $10 \, \mathrm{m \, s^{-1}}$ will provide a stress of $0.18 \, \mathrm{N \, m^{-2}}$. Over land, the drag coefficient varies according to the roughness of the surface. The surface stress on the ocean produces a surface wave field which, in turn, produces vertical motions which mix the surface momentum down into the ocean. These turbulent regions of atmosphere and ocean are known as the boundary layers, and they extend to a height of about 1000 m above the Earth's surface and to a depth of approximately 50 m in the ocean. In this section the effects of the boundary layers on flow will be considered. The two boundary layers are different because, in the atmosphere, the boundary layer is removing momentum from the flow, and it is therefore acting to retard the flow of the free atmosphere. However, in the ocean the boundary layer is the main source of momentum to drive not only the surface circulation, but also the deeper geostrophic circulation.

First, consider the influence of a frictional force on the geostrophic wind. Figure 7.15 shows the balance of forces between the pressure-gradient force, the Coriolis force and a frictional force. Basically, the frictional force opposes the motion and reduces the wind to below its geostrophic value. The Coriolis force is reduced and, in order to balance the pressure gradient, the wind must rotate towards low pressure to allow a component of the frictional force to oppose the pressure gradient. The angle of rotation is typically 30° over a land surface and 10–15° over the ocean. A component of the flow is directed down the pressure gradient, and therefore mass is transferred between high- and low-pressure systems in these frictional boundary layers. By this process, friction acts to reduce the horizontal pressure gradients and to spin down circulation systems in the atmosphere. An extra-tropical cyclone will take about six days to dissipate its kinetic

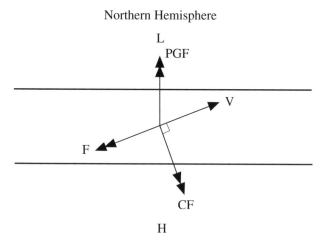

Figure 7.15 Balance of forces in the atmospheric boundary layer for a steady flow, *V*. PGF represents the pressure-gradient force, CF is the Coriolis force and F is the frictional force

energy through the surface boundary layer. It is also noted that the frictional layer will, in turn, produce surface convergence and upward motion in a low-pressure system, and surface divergence and downward motion in a high-pressure system. These circulations are known as secondary circulations. However, it must be pointed out that vertical motion in large-scale and synoptic systems is predominantly induced by the flow in the upper troposphere rather than by frictional boundary layers.

A second example of the effect of friction on motion is that in the surface ocean boundary layer. Consider a region where the horizontal pressure gradient is small and a balance of forces between the surface wind stress, the Coriolis force and a frictional force below the surface can be achieved. Here the wind stress will initially drive a current downwind but, because of the Coriolis force, which acts perpendicular to the direction of motion, there will be a clockwise rotation of the flow in the Northern Hemisphere (and an anticlockwise rotation in the Southern Hemisphere). For steady flow, frictional drag by the water below the surface layer is necessary to produce a three-way balance of forces. This clockwise rotation of the surface current with respect to the wind stress was first noted by Nansen, when he observed that pack ice drifted at an angle between 20 and 40° to the right of the wind during the Arctic Ocean expedition (1893–96) on board the research vessel *Fram*. Consideration of the balance of forces below the surface shows that the surface-layer flow will induce a stress on the layer below which, in turn, will produce a further clockwise rotation of the flow. This rotation of the flow with depth was first theorized by Ekman (1905), and it is now known as the Ekman spiral. Below about 50 m these Ekman currents are generally weak. Despite a large number of attempts to observe the Ekman spiral in the ocean, only a handful of measurements have shown its existence and these were taken under pack ice. There are probably two reasons for this difficulty in actually observing Ekman spirals. First, a fluctuating wind would induce strong inertial oscillations in the surface flow. Secondly, the surface layer is often very well mixed by vertical motion as the result of surface waves, surface convection and direct wind mixing.

The effect of the ocean surface boundary on the deeper ocean is of crucial importance in the explanation of the climatological current systems above the main thermocline. The Ekman balance between the tangential stress, τ, and the Coriolis force is given by:

$$-fv = -\frac{1}{\rho}\frac{\partial \tau_x}{\partial z} \tag{7.41a}$$

and

$$fu = \frac{1}{\rho}\frac{\partial \tau_y}{\partial z} \tag{7.41b}$$

Integrating equation 7.41 between the surface and the bottom of the frictional layer, h, yields:

$$-\int_{-h}^{0} fv \; dz = \frac{\tau_x}{\rho}$$

or

$$-\int_{-h}^{0} \rho v \; dz = \frac{\tau_x}{f} \tag{7.42a}$$

where τ_x and τ_y are the surface components of the wind stress. Similarly from equation 7.41:

$$\int_{-h}^{0} fu \; dz = \frac{\tau_y}{\rho}$$

or

$$\int_{-h}^{0} \rho u \; dz = \frac{\tau_y}{f} \tag{7.42b}$$

The left-hand sides of equations 7.42a and b are the mass transports in the frictional Ekman layer which are determined by the direction of the stress. The mass transports are perpendicular to the direction of the stress and, in the Northern Hemisphere, they act to the right of the wind stress. Figure 7.16 shows the effect of an anticyclonic and a cyclonic atmospheric circulation on the mass transport in the surface layer. In the anticyclonic case, the mass transport is directed towards the centre of the circulation and, because there can be no net accumulation of mass, mass must be removed from the surface by downward vertical motion. The downwelling, in turn, tends to bow down the thermocline in the central region of the anticyclonic

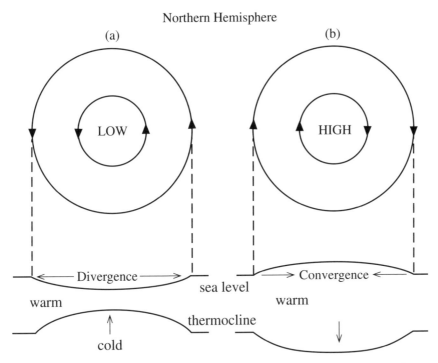

Figure 7.16   Ekman transport and vertical motion in the surface layer induced by cyclonic wind stress (a) and anticyclonic wind stress (b) in the Northern Hemisphere

circulation and the accumulation of warm, less dense water will produce a region of high dynamic topography. By the geostrophic equation (equation 7.29), this will cause a surface anticyclonic circulation relative to the deep water. In the cyclonic case, surface divergence will cause upwelling of the thermocline and so produce a central region of relatively dense, cold water which will induce a surface cyclonic flow relative to deep water.

Figure 6.14 shows a north–south section of the North Atlantic Ocean. The downwelling region is evident in the sub-tropical zone as a result of the semi-permanent atmospheric anticyclonic flow and it is responsible for the presence of relatively less dense, warm water in the deeper layers of the sub-tropical gyre. The effect of the sub-polar atmospheric cyclonic circulation is not evident because strong surface cooling and convection tend to weaken the vertical temperature gradients. On a smaller scale, tropical cyclones can have a dramatic influence on sea surface temperatures as the result of strong upwelling induced by the cyclonic circulation in the rear of the storm path.

7.6   Equatorial winds and currents

For both the atmosphere and ocean, the equatorial region is a dynamically special one. In the atmosphere the winds are both steadier and weaker than the winds at higher latitudes. Synoptic variability of the equatorial atmosphere is less than in middle latitudes, although the seasonal variations associated with the monsoonal circulations are a very significant factor. In the Pacific Ocean, interannual variations of the equatorial tropospheric and stratospheric winds are an important feature of its variability. Ocean currents tend to be stronger in equatorial regions than at higher latitudes and they are generally orientated in an east–west direction. They also respond more quickly to wind variations, as seen in the Indian Ocean where current reversals occur in response to the seasonal surface-wind circulation.

These features of the equatorial region can be related to the reduction in the Coriolis force close to the equator and to its absence on the equator. At 4º latitude, the Coriolis parameter ($2\Omega \sin \phi$) is an order of magnitude less than at 45º latitude. The geostrophic and the Ekman balances, therefore, will no longer be good approximations for winds and currents close to the equator. The question then arises as to how close to the equator can one go before the geostrophic wind equation is of no further use? Generally, it is not acceptable to apply the approximation to large-scale atmospheric flows within about 15º of the equator and to ocean flows within about 4º of the equator, for reasons which are too complex to be discussed in an introductory text. However, the above limitations do not imply that the Coriolis acceleration can be neglected. Rather, it is that other terms in the equations of motion become increasingly important.

Figure 7.17 shows a theoretical calculation of the surface mass transport caused by a steady westward wind stress in a region spanning the equator. On the equator there is no Coriolis force, and therefore a current transport in the direction of the wind stress is induced. This current will continue to accelerate until a frictional balance between the wind stress and the bottom-layer stress is

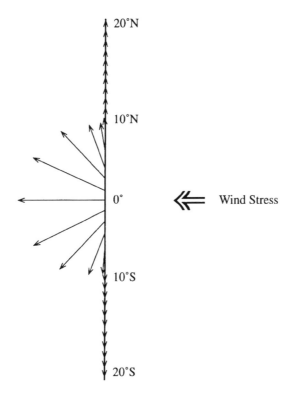

Figure 7.17   Surface transport in the equatorial ocean, caused by a westward wind stress

achieved. Within 1° of the equator the Coriolis force begins to have an effect, and it turns the current to the right in the Northern Hemisphere and to the left in the Southern Hemisphere. Therefore, there is a divergence of surface flow from the equator which, by the conservation of mass, must be balanced by an upward motion of water into the surface layer. This upwelling is of the order of $1\,\mathrm{m\,day^{-1}}$ and it causes a lifting of the thermocline. The equatorial thermocline usually lies between 100 and 200 m depth but, in local areas where upwelling is strong, it may be less than 50 m below the surface. At about 10° latitude the surface transport is nearly perpendicular to the wind stress, in agreement with the Ekman balance (equation 7.42). In regions where the surface-wind stress is directed eastward, such as in the northern Indian Ocean during the south-west monsoon, then the surface transport will converge onto the equator. In this case, downward motion out of the surface layer will occur and this will depress the equatorial thermocline.

The meridional tilt of the thermocline, caused by equatorial upwelling, will induce a meridional pressure gradient with high dynamic height polewards of the equator. A few degrees of latitude from the equator, the flow will be close to geostrophic, and therefore a westward flow within the thermocline will occur. It is this meridional pressure gradient which is responsible for the north and south equatorial currents which are found in all ocean basins, except for the northern Indian Ocean during the south-west monsoon. In this latter case, because of downwelling on the equator, the meridional pressure gradient is reversed and eastward equatorial currents occur. A downwelling region in the eastern equatorial Pacific and Atlantic Oceans, near 10°N, results in a similar reversal of the pressure gradient and induces a narrow geostrophic countercurrent towards the east. These examples show the intimate relationship between the surface wind, upwelling and pressure gradients, and the production of strong, zonal equatorial currents.

Along the equator the primary balance of forces in the upper ocean is between the pressure-gradient force and the surface-wind stress. Figure 7.18a shows the zonal pressure gradient on the equator in the Atlantic Ocean for February and September. In February, the surface winds are generally weak westward winds and there is only a small eastward pressure gradient. However, in September, very strong westward trade winds extend across the equator and a large eastward pressure gradient is required for balance. The highest dynamic heights occur in the west and here the thermocline is depressed, relative to the eastern part of the region. Figure 7.18b shows the depth of the 23°C isotherm, which is indicative of the depth of the thermocline and which changes its slope in response to the surface-wind stress. The upward slope of the thermocline towards the east accounts for the frequent appearance of cold, upwelling water in the eastern equatorial Atlantic and Pacific Oceans. In the Indian Ocean during the south-west monsoon, the zonal tilt of the thermocline is reversed.

If the horizontal pressure gradient exactly balanced the surface-wind stress, there would be no motion along the equator. However, the balance of forces is not exact and the currents respond quickly to the imbalance. Below the surface layer, unimpeded by the wind stress, the eastward pressure gradient in the thermocline drives an eastward

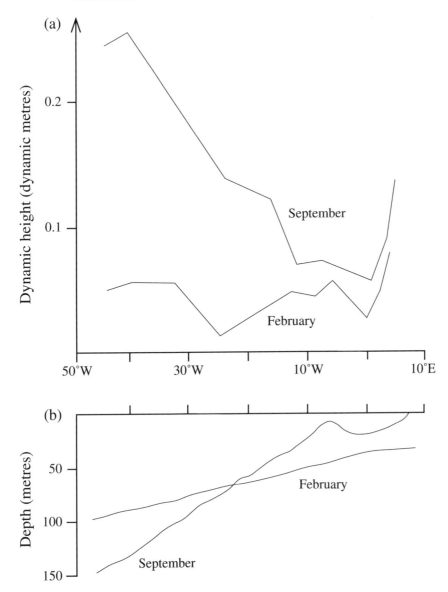

Figure 7.18 (a) Zonal pressure gradient in the equatorial Atlantic in February and September. (b) Depth of 23°C isotherm (equivalent to depth of the thermocline) in February and September

undercurrent along the equator. In the Pacific Ocean, the equatorial undercurrent is 14 000 km long, but only about 400 km wide. It is confined to within 2° of the equator and has a maximum velocity of $2\,\mathrm{m\,s^{-1}}$ close to its core in the thermocline. In both the Pacific and Atlantic Oceans, the undercurrent slopes upwards towards the east and it can, occasionally, surface in the eastern basins.

The following calculation will give an indication of the maximum eastward acceleration which can occur as a result of the imbalance of the observed pressure gradient with the wind stress. From Figure 7.18a, it can be seen that the maximum difference in dynamic height between 40°W and 0°E in the Atlantic Ocean is 0.2 dynamic metres. The horizontal distance is 40° longitude, or 4.4×10^6 m. Therefore, the pressure-gradient acceleration is $(0.2 \times 10\,\mathrm{m^2\,s^{-2}})/(4.4 \times 10^6\mathrm{t\,m})$, or $4.5 \times 10^{-7}\,\mathrm{m\,s^{-2}}$. The velocity, u, after constant acceleration, a, through a distance, s, is given by $u^2 = 2as$. Assuming uniform acceleration across the Atlantic Ocean:

$$u = \sqrt{(2 \times 4.5 \times 10^{-7} \times 4.4 \times 10^6)}\,\mathrm{m\,s^{-1}}$$
$$= 2.0\,\mathrm{m\,s^{-1}}$$

This simple calculation therefore gives a reasonable estimate for the maximum observed velocity of the equatorial undercurrent. In practice, frictional turbulence with the deeper ocean and horizontal mixing associated with meanders in the undercurrent will tend to reduce the observed eastward acceleration below that assumed in the above calculation.

Some features of the equatorial atmosphere will now be considered. One of the dominant seasonal variations is the Asian monsoonal circulation which, during the summer, causes a surface atmospheric flow from the southern Indian Ocean to the Asian continent. This flow can be intense, with winds of up to $20\,\mathrm{m\,s^{-1}}$ occurring off Somali. Figure 7.19 is a simplified map of surface pressure for the south-west monsoons and shows that the main feature is the curvature of the winds as they cross the equator. There is a pressure gradient directed from the sub-tropical high pressure in the southern Indian Ocean towards the Indian sub-continent. In the Northern Hemisphere, away from the equator, the geostrophic wind is eastwards, whilst in the Southern Hemisphere it is westwards. However, because of surface friction, the westward flow will rotate clockwise as shown. Consider a parcel of air in the Southern Hemisphere and moving towards the equator. As the parcel approaches the equator, the deflecting Coriolis

LOW PRESSURE

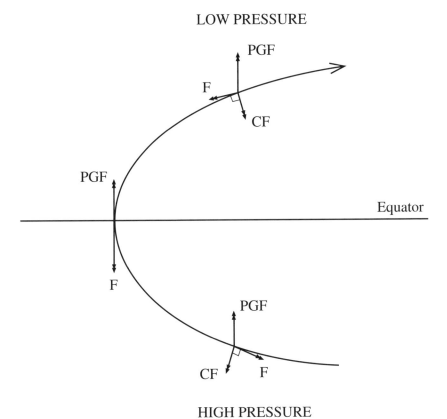

Equator

HIGH PRESSURE

Figure 7.19   Balance of forces in a monsoonal flow (summer monsoon) across the equator. The horizontal pressure-gradient force is PGF, the Coriolis force is CF and the frictional force is F

force is weak and the parcel will accelerate northwards down the pressure gradient. Only friction would be able to reduce the acceleration at this stage. Once the parcel crosses the equator into the Northern Hemisphere, it will begin to feel the effect of the Coriolis force now acting to the right and it will gradually be deflected towards the east.

A second feature of the monsoonal circulation is the appearance of an intense, low-level jet stream, as shown in Figure 7.20a. It appears off the coast of East Africa during the south-west monsoon. This jet stream, unlike those in the middle latitudes, is directed from the south

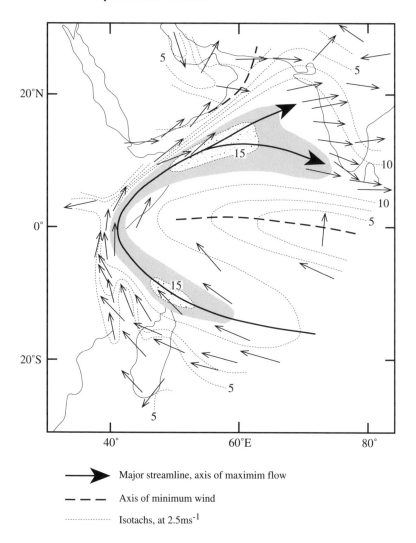

20°N

0°

20°S

40°   60°E   80°

➤ Major streamline, axis of maximim flow

— — — Axis of minimum wind

·············· Isotachs, at 2.5ms⁻¹

Figure 7.20 (a) East African jet in the atmosphere: monthly mean air flow at 1 km in July. Wind speeds at 2.5 m s⁻¹ intervals. (b) Surface currents in the Indian Ocean during the summer monsoon. The intense flow adjacent to the African coast is the Somali jet. The numbers on the dotted lines indicate wind speeds (isotachs) in m s⁻¹. Crown ©

to the north across the equator, and velocities of up to $25 \, \text{m s}^{-1}$ are reached between a height of 1 and 2 km. It was discovered by Findlater, a British meteorologist, who mapped the complete jet stream after receiving reports of very strong, anomalous winds from civil aircraft operating from East Africa. The formation of this jet stream appears to be analogous to that of the western boundary currents in the ocean. In the atmospheric case, the East Africa tablelands provide a convenient western boundary to the low-level monsoon flow. The relationship between the low-level jet stream, the Somali current and the intense upwelling off the Somali coast is a major area of research for both physical oceanographers and meteorologists. In 1979, during the First Global GARP Experiment (FGGE), a detailed oceanographic and atmospheric study of the summer Indian monsoon was undertaken. Observations showed that the atmospheric jet was exceptionally strong and surface stresses of over $0.4\,N\,m^{-2}$ were recorded near the Somali coast at 10ºN in June and July. Sea surface temperatures near the coast dropped from 27ºC at the beginning of June (i.e. the time of the monsoon onset) to 18ºC by the end of the month as the result of intense upwelling. Measurements of surface currents by ships' drift in August showed north-eastward currents of between 3 and $4\,m\,s^{-1}$ (Figure 7.20b).

An important feature of the equatorial ocean circulation is its fast response to seasonal variations in wind. In the Atlantic and Indian Oceans, the response time is less than one month, whilst for the Pacific Ocean, because of its large zonal scale, the response time is between two and three months. In the Indian Ocean, as discussed above, the Somali current responds to the onset of the south-west monsoon within two to three weeks. In contrast, the large sub-tropical gyres in the middle latitudes respond on time scales of years to decades, and they therefore show little seasonal variation. In this respect, the atmosphere and ocean circulations in low latitudes respond on similar time scales (i.e. seasonal and interannual), and therefore it is in these regions that strongly coupled interactions can take place. This will be discussed in more detail in Chapter 10.

8

Waves and Tides

8.1 The spectrum of surface waves

Before discussing the spectrum of surface waves, it is necessary to introduce a few definitions. First, the wave height, H, is the vertical distance between the wave trough and the wave crest, and it is twice the amplitude, a, of the wave. The wave period, T, is the time interval between the passage of two consecutive wave crests, or troughs, at a fixed point. The wavelength, L, is the distance between two wave crests measured in the direction of propagation. The individual wave speed, c, is given by $c = L/T$, the frequency, σ, by $\sigma = 2\pi/T$ and the wavenumber, k, by $2\pi/L$. The wave, or phase speed, is given by $c = \sigma/k$. The speed of energy propagation of a wave is related to the group velocity of the waves, c_g, where $c_g = \partial\sigma/\partial k$. If the wave speed, c, is not dependent on the wavenumber of the wave, then the wave is said to be non-dispersive and $c = c_g$.

Figure 8.1 shows the energy spectrum of surface waves in the ocean for periods ranging from 0.01 to 10^5 s. The highest-frequency components of the spectrum are the capillary waves which have wavelengths of a few centimetres and periods of less than 0.1 s. They are generated, almost instantaneously, by gusts of wind which produce the familiar 'cats paws' ripple pattern on an otherwise smooth water surface. They are also dissipated in less than 30 s by viscosity. The restoring forces for these capillary waves are, in general, both gravity and surface tension, and the wave speed, c, is given by:

$$c = \sqrt{\left(\frac{gL}{2\pi} + \frac{2\pi S}{\rho L}\right)} \qquad (8.1)$$

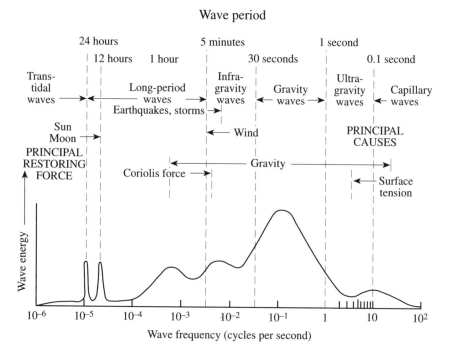

Figure 8.1 Schematic representation of the relative amounts of energy in waves of different periods

where S is the surface tension of water, which is typically $74 \times 10^{-3}\,\mathrm{N\,m^{-1}}$. It can be shown that the minimum wave velocity, c_m, occurs when $gL_\mathrm{m}/2\pi = 2\pi S/\rho L_\mathrm{m}$, i.e. when $L_\mathrm{m} = 2\pi\sqrt{(S/\rho g)}$, where L_m is the wavelength corresponding to the minimum wave velocity. For sea water, $L_\mathrm{m} = 1.7\,\mathrm{cm}$ and $c_\mathrm{m} = 23\,\mathrm{cm\,s^{-1}}$. At wavelengths shorter than L_m, the surface tension in equation 8.1 becomes increasingly dominant and c increases as L decreases. For wavelengths longer than L_m, gravity becomes more important than surface tension. In this case, c increases as L increases. An interesting peculiarity of these waves is that the group velocity is greater than the phase velocity and so a packet of capillary waves will move more quickly than the individual waves. The profile of a capillary wave is quite unlike that of a gravity wave in that it has a flat crest and a sharp trough. Capillary waves, because of their dependence on surface tension, are strongly affected by surface films of oil or other organic matter and, because the absence of capillary waves on the surface changes the reflection, surface slicks are easily seen. More recently, it has been shown that, because gravity–capillary waves are a sensitive indicator of wind speed, the magnitude of the reflection of microwave radiation (λ = 2.2 cm) can give a quantitative measurement of wind speed. The instrument, known as a scatterometer, has been shown to give estimates of wind speed accurate to within 1 or $2\,\mathrm{m\,s^{-1}}$ when flow on the SEASAT satellite in 1978.

The most significant energy band in the spectrum is that of wind waves and swell, which have periods between 1 and 30 s. In this frequency band, gravity is the only restoring force, and therefore these waves are true gravity waves generated by a turbulent surface-wind stress. The particles in these waves execute circular orbits and are not transported by the waves. The radius of the orbits of the particles decreases with depth according to the relationship $r = a \exp[-(2\pi/L)z]$, where a is the amplitude of the wave. At a depth of half the wavelength of the surface wave, the particles' orbit radius is $a/23$. For example, a surface wave having a 1 m amplitude and a 10 m wavelength would have a 4 cm amplitude of particle motion at a depth of 5 m. Therefore, wave motions can be seen to be quickly damped with depth. It is noted that the profile of the wave is not exactly sinusoidal. The wave crests are sharper and the wave troughs are flatter than in a true sinusoidal wave. For the larger-amplitude waves this effect becomes more marked, with wave crests reaching a limiting angle of 120°, corresponding to a wave slope of about 1 in 7. At this stage, wave breaking occurs.

For a wave whose amplitude is small compared with wavelength, the wave speed, c, is given by:

$$c^2 = \frac{gL}{2\pi} \tanh\left(\frac{2\pi h}{L}\right) \tag{8.2}$$

where h is the depth of the water. Such waves are known as linear gravity waves and *for waves whose wavelength is smaller than twice the depth of the water* (i.e. $h/L > 0.5$), then $\tanh(2\pi h/L) \simeq 1$ and $c^2 = gL/2\pi$. Since $c = L/T$, it can be easily shown that:

$$c = \frac{g}{2\pi}T \quad \text{and} \quad L = \frac{g}{2\pi}T^2 \tag{8.3}$$

Table 8.1 shows numerical values of c, L and T for typical ocean waves. It is noted that long waves have higher wave velocities than short

Table 8.1   Characteristics of wind waves and swell, derived from equation 8.3

	Wind waves	Swell	Long swell
Period, T (s)	1–5	10	20
Wavelength, L (m)	1.6–39	156	624
Phase velocity, c $(m\,s^{-1})$	1.6–7.8	15.6	31.2
Group velocity, c_g $(m\,s^{-1})$	0.8–3.9	7.8	15.6

waves and that the wavelength is proportional to the square of the period. These gravity waves are dispersive since their speed is dependent on wavelength, and it would therefore be expected that the group velocity of a packet of wind waves would be different from the phase speed. Since $c = \sigma/k$ and $T = 2\pi/\sigma$, it follows that $\sigma^2 = gk$ or $\sigma = \sqrt{(gk)}$. Differentiating:

$$c_g = \frac{\partial \sigma}{\partial k} = \frac{1}{2}\sqrt{(g/k)} = \frac{1}{2}c \qquad (8.4)$$

Therefore, the group velocity is equal to half of the wave speed, c. This deduction can be easily verified by studying the propagation of a group of waves produced by a ship or a boat. The individual wave crests will be seen to move towards the observer through the group of higher waves and then they will be seen to die out.

As mentioned earlier, apart from tides, the largest energy density in the surface-wave spectrum is associated with wind waves and swell. A simple example illustrates the typical energy in ocean swells. The energy density of a sinusoidal wave is $\rho g a^2/2$ or $\rho g H^2/8$, and therefore a wave with a height of 2 m would have an energy density per unit area of $5 \times 10^3\,J\,m^{-2}$. If it is assumed that the ocean swell has a 10 s period, then its group velocity is $\frac{1}{2} \times 15.6\,m\,s^{-1}$ or $7.8\,m\,s^{-1}$. The energy flux per unit time, impinging on a straight coastline, is the product of the energy density and the group velocity. In the present example, this flux is $7.8\,m\,s^{-1} \times 5 \times 10^3\,J\,m^{-2}$, or approximately 4 kW for each metre of coastline. Most of this energy is dissipated into heat in the wave-breaking zone.

In water depth which is small compared with the wavelength (i.e. $h/L < 0.5$), $\tanh(2\pi h/L) \simeq (2\pi h/L)$ and,   from   equation   8.2,

$c = (gh)^{\frac{1}{2}}$. Unlike the ordinary gravity waves for which $h/L > 0.5$, the wave speed is independent of wavelength and the group velocity is equal to the phase speed. Such waves are known as long waves, and particle motions in long waves are horizontal and do not decay with depth. Long waves appear in the spectrum between swell and tides, and they have periods between 30 s and 12 hours.

An extreme example of a long wave is the tsunami, or seismic wave, which can produce locally high-energy densities. These seismic waves are caused by submarine volcanoes, earthquakes or landslides, and they propagate quickly at speeds of about 200 m s^{-1} in deep ocean water (i.e. 4 km depth) and at about 30 m s^{-1} on continental shelves where the ocean is less than 100 m deep. Their wavelengths are very long and, for a tsunami having a wave period of 10^3 s, the wavelength in deep ocean water would be approximately 200 km, and therefore the wave would be difficult to detect. However, in shallow water, where it propagates more slowly and where the wave energy is concentrated into a smaller depth, the amplitude of the wave would grow catastrophically.

Other examples of long waves include 'surf beat', which has a period of a few minutes and can be observed as a long-period oscillation in water level on a beach. It is caused by a long wave travelling longitudinally between the surf zone and the shore. A second example is the shelf waves which travel along the 'shelf break' at the edge of the continental shelf. They have periods between 30 min and two hours, and can travel for hundreds or even thousands of kilometres along the shelf break.

The extreme low-frequency long waves are the tides which are forced by gravitational attraction between the Earth, the Moon and the Sun. The tides will be discussed in Sections 8.4 and 8.5.

8.2 Wind waves and swell

In the previous section, a number of formulae have been briefly discussed which have been derived from elegant theories of surface waves developed by mathematicians over the last two centuries. It has also been seen that they make available practical tools with which to calculate relationships between wave speed, period and wavelength. However, these formulae and theories give very little insight into how to describe the complicated surface wave patterns observed in nature, or into the understanding of how waves are generated and how they decay. It was the need to answer such questions that gave rise to the first quantitative study of ocean waves in the early 1940s. At this time, military planners required predictions of wave heights and swell for seaborne operations, and in particular for beach landings.

During this early period, a number of different methods were devised to measure the height of the sea surface covering a frequency range of 1 to 50 s. One of the most successful instruments was the underwater pressure recorder which could be located either on the bottom, in water less than 100 m in depth, or under the water line of a ship. The pressure fluctuations, a, at depth, h, can be used in conjunction with wave theory to give the amplitude of the variation of the sea surface, using the following relationship:

$$a_0 = a_h \cosh{(kh)} \qquad (8.5)$$

where a_0 is the amplitude of the sea surface, k is the wavenumber and a_h is the amplitude at depth h. The signals can be recorded on magnetic tape in the ship or at the shore base. This type of instrument tends to filter out high-frequency fluctuations. A second method involves the measurement of the vertical displacement of a ship or buoy using an accelerometer. Many of the measurements of ocean waves have been obtained from ships equipped with pressure recorders on the hull and accelerometers, the latter being used to correct for the ship's motion. Other methods include the use of inverted echo-sounders, either moored or on the bottom, to obtain surface displacement, and the use of radars and lasers, either on fixed platforms or on aircraft and satellites. All of these methods have their own advantages and disadvantages, and in the end it depends very much on the question being asked as to what system of measurement is adopted.

A typical trace from a bottom-pressure wave recorder is shown in Figure 8.2. It confirms previous suspicions that the sea surface is, in general, irregular and does not undergo simple sinusoidal variations. It is clearly not possible to define a single wave height or a single wave period for this record. In order to analyse such a record, it is necessary to decompose it into its component frequencies and amplitudes, using Fourier analysis. This is done by dividing the record into blocks of

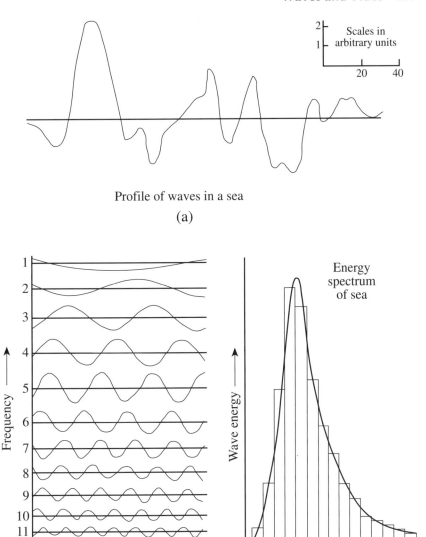

Profile of waves in a sea

(a)

(b)   (c)

Figure 8.2   An observed profile of waves in the sea (a). Such a complicated wave pattern can be described as consisting of many different sets of sine waves (b), all of them superimposed. The lower-frequency waves contain more energy than the higher frequency ones (c). The energy in a wave is proportional to the square of its height equal periods, say 4 min, and then calculating the Fourier coefficients for each block of the sea-level record, Z_j. If the sampling rate is 2 s, then 120 data points are obtained for each block which, in turn, will give Fourier coefficients, H_n, for 59 discrete frequencies, thus:

$$H_n = \frac{1}{60} \sum_{j=1}^{120} Z_j \exp\left(i \frac{2\pi j n}{120} \right) \text{ for } n = 1, 59 \qquad (8.6)$$

The energy, E_n, for each frequency, n, is then proportional to the modulus of the Fourier coefficients, $|H_n|^2$. In order to obtain a statistically valid sample, the procedure is repeated for subsequent blocks of data and the average energy, E_n, is calculated for each frequency. The average energy is then plotted against frequency to produce an energy or power spectrum such as the one shown in Figure 8.2. First, this type of analysis shows that, in general, there is a continuous frequency distribution of waves with the maximum energy at the frequency of the dominant wave, F_0. Secondly, the distribution is not symmetrical in that it has a more rapid reduction towards lower frequencies. Finally, the total energy of the wave field is proportional to the area under the curve.

Two spectral examples will now be considered. Figure 8.3 shows a series of wave spectra observed off the west German coast when the wind was blowing steadily from the land. From common experience, one expects the height of the waves, and therefore their energy, to increase with the distance offshore, or the fetch. In the example, the areas under the curves increase markedly with the fetch. The spectra also show that the dominant wave frequency decreases with fetch, from 0.4 to 0.2 cycles s^{-1}. This corresponds to an increase in wave period from 2.5 to 5 s. Secondly, as the fetch increases, the spectra become narrower and this implies that more of the wave energy becomes concentrated in fewer frequencies. At each position the wave spectra are in a steady state. This means that, although energy is being added to the waves by the wind, energy is being dissipated at about the same rate. Such spectra are known as 'saturated' wave spectra, and their shape, dominant wave frequency and total energy can be related to two parameters, namely, (i) the fetch and (ii) the wind stress.

The second example is illustrated by Figure 8.4, which shows the change of shape of spectra with time under steady wind conditions, i.e.

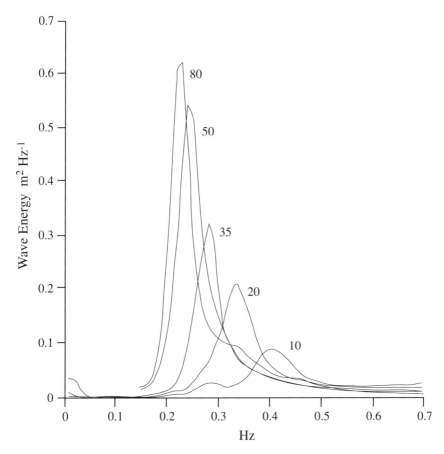

Figure 8.3   Evolution of wave spectrum with fetch for offshore winds from 11.00–12.00 hours, 15 September 1968. The fetch, in kilometres, is shown adjacent to each spectrum

prior to the attainment of a steady state. The high-frequency waves become saturated within a relatively short period of time, but the lower-frequency waves continue to grow during the three-hour period. Unlike the previous example, the dominant wave frequency remains relatively constant during the period, but the spectra exhibit a similar narrowing with time. The time taken to reach a saturated spectrum depends upon the wind speed. For relatively low wind speeds of $5\,\mathrm{m\,s^{-1}}$, saturation will be attained in about two to three hours, whilst for higher wind speeds of, say, $15\,\mathrm{m\,s^{-1}}$, it may take 24 hours. As the fetch increases beyond a critical value, the spectra become independent

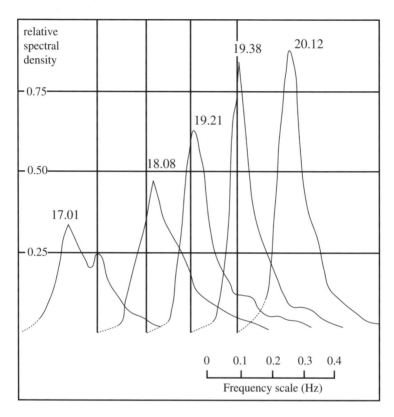

Figure 8.4 Evolution of wave spectra with time on 28 October 1976, between 17.00 and 20.00 hours. The wind increased from $3\,\mathrm{m\,s^{-1}}$ to $10\,\mathrm{m\,s^{-1}}$ from 15.00 to 16.00 hours and then was constant for the remaining period

of fetch and they can be directly related to the wind velocity. At low wind speeds of $5\,\mathrm{m\,s^{-1}}$, this will occur at a distance of approximately 10 km offshore but, for a higher wind speed of $15\,\mathrm{m\,s^{-1}}$, a fetch of more than 500 km is required.

Provided that the wave spectra are in equilibrium with the wind and they are not limited by fetch from the coast, then they can be used to relate the statistics of the wave field to the wind speed. The total energy of the wave field, E_T, at a given wind speed is equal to the area under the spectral curve, and the quantity $\sqrt{E_T}$ is a measure of the root-mean-square height of the waves. If the significant wave height, H_s, is defined as the mean wave height of the one-third highest (peak to trough) waves, then it has been shown that:

$$H_s = 4.0\sqrt{E_T}$$

Spectral decomposition is, therefore, a powerful tool for the presentation of large quantities of wave data in a succinct manner and for obtaining empirical formulae which can be used for wave prediction. Wave spectra also have considerable scientific value in the understanding of how waves are generated and dissipated. The two previous examples show how wave fields grow with time and change with offshore distance, and thus give important clues to the actual mechanisms by which waves are generated. Two mechanisms have been proposed for the development of waves by wind. In the first mechanism, it is assumed that a wind blowing across a water surface will produce surface pressure fluctuations as a result of turbulence in the boundary layer. This will in turn cause small waves to form. If some of the waves move at a similar speed to these pressure fluctuations, then the pressure field will reinforce the wave field and the waves will grow. Theoretical analyses have shown that the wave energy will grow linearly with time. The important aspect of this theory is the assumption that waves do not affect the pressure fluctuations in the atmosphere and it is, therefore, most applicable to the initial generation of waves when amplitudes are small. The second mechanism assumes that the surface waves will affect the pressure fluctuations in the atmosphere in such a way that surface pressure fluctuations and waves grow with time. If the flow over the waves is smooth, then the pressure distribution cannot transmit energy into the waves but, if the flow is turbulent, the pressure distribution can come into phase with the vertical velocity of the sea surface and allow a rapid energy exchange. This causes the waves to grow exponentially with time.

These two mechanisms fail to explain two features of the spectrum; namely, the shift of the spectral peak to low frequencies with time and the development of a saturated spectrum. The mechanisms have recently been studied in the JONSWAP experiment in the North Sea. Figure 8.5 shows the wave energy transfer measured for a developing wave spectrum during this experiment. It may be seen that most of the wind energy goes into the spectrum over a broad band of frequencies, whilst the dissipation is concentrated at the high-frequency end of the spectrum. In order to achieve a balance between energy sources and sinks, energy has to be transferred to high-frequency waves. This has to be done by interaction between waves of different frequencies. A

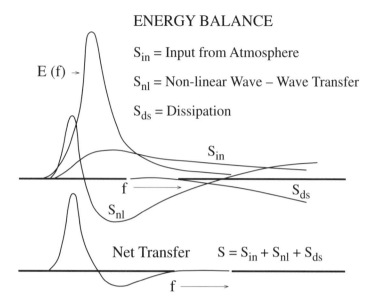

ENERGY BALANCE

S_{in} = Input from Atmosphere

S_{nl} = Non-linear Wave – Wave Transfer

S_{ds} = Dissipation

Net Transfer $S = S_{in} + S_{nl} + S_{ds}$

Figure 8.5   Schematic energy balance of the wave spectrum

qualitative example of the process is to consider two sinusoidal waves with frequencies σ_1 and σ_2, and amplitudes a_1 and a_2. Frequency σ_1 is greater than frequency σ_2. The interaction of these two sine waves is given by $a_1 a_2 \sin \sigma_1 t \sin \sigma_2 t$. Using sum and difference formulae, the following terms are obtained:

$$\cos(\sigma_1 - \sigma_2)t \text{ and } \cos(\sigma_1 + \sigma_2)t$$

Therefore, the interaction between the two original waves has produced one wave with a higher frequency, $\sigma_1 + \sigma_2$, than either of the original waves and a second wave with a lower frequency. Hence, by this non-linear interaction, energy is transferred to both high and low frequencies. In the wave spectrum, the energy which is transferred to the higher frequency is dissipated by wave breaking (white caps and, ultimately, by viscosity). The energy which is transferred to the low frequencies is responsible for the spectral shift to low frequencies during a developing sea, as shown in Figure 8.3.

The low-frequency limit of the wave spectrum is the swell which propagates away from the region of wave generation and is dissipated only slowly. The low-frequency swell will propagate with a group velocity greater than that of wind-generated waves and therefore, at a distance from the storm, the lowest-frequency swell waves will be the first to arrive. These will be followed by progressively slower, higher-frequency waves. Consider a wave recording station at a distance L from a storm. From equations 8.3 and 8.4:

$$c_g = \frac{g}{2\sigma}$$

where σ is the wave frequency. The time, t, taken for the swell to travel a distance L is:

$$t = \frac{2\sigma}{g}L \qquad (8.7)$$

Therefore, the frequency of the waves arriving at the recording station will increase with time at a rate proportional to L. From a graph of wave frequency against time, the gradient will give the distance travelled and the intercept will give the time of origin of the waves. If the direction of the swell is also measured, or if wave information is available from an additional station, then the position of the storm can be ascertained. Low-frequency swell can travel for very long distances because of the small attenuation and the swell will propagate along great circle routes, as shown in Figure 8.6. This map shows the possible great circle routes to the Island of St Helena at 16°S. This island has been the subject of much interest in wave research because of the arrival of exceptionally large rollers, or swell, between December and March. From work by Cartwright *et al.* (1977), it has been shown that this swell may originate from winter storms in the Newfoundland region of the Atlantic Ocean. From wave records taken during a large-swell event on St Helena, Cartwright showed that the origin of the swell on the great circle route was coincident with an intense storm in the Newfoundland region eight days prior to the arrival of the swell. Similar studies have shown swell propagation along great circle paths from the South Pacific to Alaska. It is noted that the group velocity of the swell can be greater than the velocity of the generating storm, and so the arrival of swell can give forewarning of a tropical cyclone or a middle-latitude depression. In February 1979 some exceptional swell waves, with a period of 18 s and significant height of 7 m, propagated

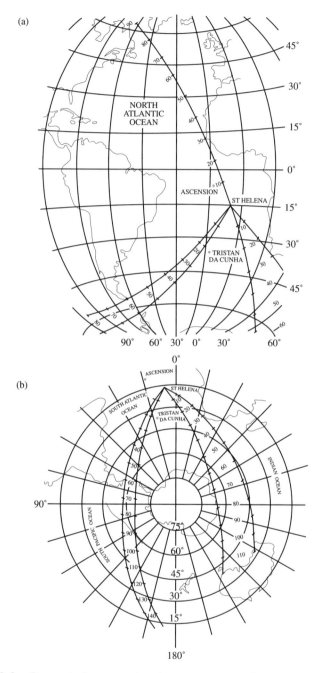

Figure 8.6   Great circle routes for the propagation of swell to the island of St Helena, from (a) the North Atlantic Ocean and (b) the Southern Ocean

along the English Channel causing considerable damage to Portland. These waves originated from an intense depression in the central Atlantic.

8.3 Long waves

The main feature that distinguishes long waves from wind waves is that their wavelength is large compared with the water depth, and therefore the waves feel the whole depth of the ocean. Water particles oscillate horizontally and are in phase over the depth of the ocean in a long wave. Only close to the bottom boundary will they show any attenuation. Provided that the elevations of the waves are small compared with the mean water depth, then they will propagate with a wave speed $\sqrt{(gh)}$.

Long waves in the ocean may be caused by large falls in surface pressure associated with tropical cyclones or by seismic activity. These waves, like swell waves, tend to follow great circle routes, and the location and time of origin of the waves can be obtained by noting the time of their arrival. By this method the origins of major submarine earthquakes have been located in the Pacific Ocean. In deep water such as the Pacific Ocean, the variations in bottom depth are relatively small compared with the total depth of the ocean, and therefore the wave speed is little affected by the bottom. However, in shallow water the bottom variations reduce the speed of the long waves considerably. As a wave approaches a shore its speed will diminish but its frequency will remain unchanged, and so the wavelength becomes shorter and the waves appear to bunch. Mathematically:

$$L = 2\pi \frac{\sqrt{(gh)}}{\sigma} \qquad (8.8)$$

The flux of energy per unit time, F, across a coast is the product of the wave-energy density and the group velocity. This flux of energy will remain constant and therefore, because of the reduction in group velocity, the wave-energy density must increase. The wave energy is proportional to the square of the wave height, and hence the wave height, H, will increase as the shore is approached. Thus:

$$F \propto \sqrt{(gh)}H^2$$

$$\therefore \; H \propto \frac{F^{\frac{1}{2}}}{(gh)^{\frac{1}{4}}} \tag{8.9}$$

If a wave has height H_1 in water of depth h_1, then equation 8.9 implies that its height H_2 in water depth h_2 is given by:

$$H_2 = H_1 \left(\frac{h_1}{h_2} \right)^{\frac{1}{4}} \tag{8.10}$$

For a tsunami which has an elevation of 1 m in water 4 km deep propagating into a coastal region having water 10 m deep:

$$H_2 = 4.47 \text{m}$$

Tsunami waves between 3 and 5 m high have been measured in coastal locations in the Pacific Ocean. It is also noted that, for a tsunami wavelength of 100 km in water 4 km deep, equation 8.8 predicts that the wavelength will decrease to 5 km in the shallow water.

For wind waves and swell a similar phenomenon occurs but, because of the increase in slope of the waves, they become unstable and break to form the surf zone. Refraction effects cause the wave front approaching at an angle to the shore to bend towards the shore.

For tsunamis, the wave frequencies are approximately 10^{-3} s^{-1}, or about 15 min. This is considerably less than the Coriolis frequency of 10^{-4} s^{-1}, and therefore these waves are not affected by the Coriolis force. However, for longer wave periods the Coriolis force has to be taken into account. Long waves forced by tides or low-frequency meteorological forcing are all considerably influenced by the Coriolis force.

Lord Kelvin first described the behaviour of a long wave in a channel constrained by the Coriolis force. Figure 8.7 shows the behaviour of the sea surface and currents in a Kelvin wave. The effect of the Coriolis force is to constrain the wave propagation to the boundary of the channel. The elevation, ξ, is given by:

$$\xi = \xi_0 \exp(-y/L) \cos(\sigma t - kx) \tag{8.11}$$

where $L = \sigma/fk = \sqrt{(gh)}/f$, f is the Coriolis parameter, σ is the wave frequency, k is the wavenumber and y is the perpendicular distance from the side of the channel.

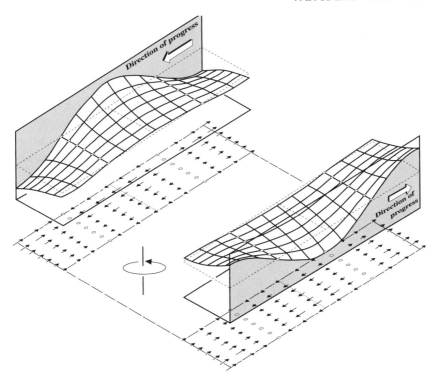

Figure 8.7   Topography of the sea surface for a Kelvin wave in a wide channel in the Northern Hemisphere. Note the wave progresses with the boundary on its right-hand side in the Northern Hemisphere

At the side of the channel, $y = 0$ and the wave has a maximum amplitude ξ_0. The amplitude decays with distance and will become negligible for distances much greater than L. For a long-wave speed of $30\,\mathrm{m\,s^{-1}}$, typical of waves on the continental shelf, $L = 30/10^{-4} \simeq 300\,\mathrm{km}$. Basically, the Kelvin wave propagates as a long wave along the boundary, but it is in geostrophic balance normal to the boundary. Therefore, the current in the x direction is in geostrophic balance with the pressure gradient caused by the slope in the sea surface, i.e.:

$$fu = -g\frac{\partial \xi}{\partial y} \tag{8.12}$$

Substituting for ξ in equation 8.12 gives:

$$u = \sqrt{(g/h)}\xi \qquad (8.13)$$

Therefore, high elevation is associated with a positive current and low elevation with a counter current. In a basin closed at one end, a Kelvin wave will propagate around the coast in an anticlockwise direction in the Northern Hemisphere and a clockwise direction in the Southern Hemisphere. Examples of these waves will be discussed with tides and storm surges in later sections of this chapter.

8.4   Internal waves

Consider two fluids having densities ρ_1 and ρ_2 in hydrostatic equilibrium. If the interface between the two fluids is moved upwards from the equilibrium position, then an element of denser fluid ρ_2 will be subject to a downward buoyancy force proportional to $(\rho_2 - \rho_1)g$. If the two fluids are the ocean and the atmosphere, then ρ_2 is approximately $1025\,\mathrm{kg\,m^{-3}}$ and ρ_1 is $1.25\,\mathrm{kg\,m^{-3}}$. In this case the buoyancy force is, to a very good approximation, proportional to $\rho_2 g$. The fact that the ocean density is so very much larger than the atmospheric density is the reason why the density of the atmosphere is so often neglected in formulae for surface waves. However, for two fluids of similar densities, the $(\rho_2 - \rho_1)$ term, and hence the buoyancy force, will be small. For example, consider a layer of fresh water overlying a layer of ocean water (a situation commonly encountered in estuaries and fjords); then the difference in density is $25\,\mathrm{kg\,m^{-3}}$, assuming that the fresh water has a density of $1000\,\mathrm{kg\,m^{-3}}$. The restoring buoyancy force is only 2.5% of the buoyancy force for a surface gravity wave, and therefore the restoration of the interface to equilibrium will take considerably longer than for a surface wave. Internal waves, therefore, generally have longer periods and move more slowly than their surface counterparts.

When the layer of lighter water overlies a considerably deeper layer of denser water, then the propagation speed of the internal wave is:

$$c = \sqrt{\left(\frac{gh_1(\rho_2 - \rho_1)}{\rho_2}\right)} \qquad (8.14)$$

where h_1 is the depth of the upper layer. In a fjord the surface layer may be approximately 4 m deep and therefore, from equation 8.14, the internal wave speed will be $1\,\mathrm{m\,s^{-1}}$. This may be compared with the surface wave speeds in Table 8.1. Figure 8.8 shows the circulation induced by an internal wave in a two-layer fluid. It is noted that the density interface has much larger amplitude displacements than those which occur at the surface, and therefore the circulation in the upper layer becomes intensified at the crest of the internal wave in the opposite direction to that of the internal wave. A ship in a fjord will not only produce a surface bow wave, but also an internal wave, if the surface layer is shallow. Depending on the speed and length of the boat, the reversed surface circulation at the crest of the internal wave will impede the progress of the boat. This phenomenon is known as 'dead' water because boats caught in these regions will make reduced headway through the water.

A second example of internal waves includes those whose wavelengths are much longer than the total depth of the fluid.

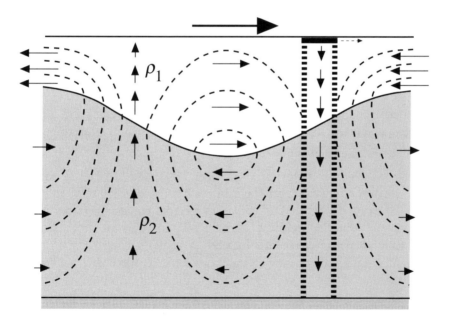

Figure 8.8 Circulations (streamlines are shown as dashes) associated with progressive internal waves on a density interface between a lighter fluid, ρ_1, and a denser fluid, ρ_2. Surface slicks form in the region of surface convergence

Such waves are termed long internal waves. Their wave speed, c, is given by:

$$c = \sqrt{\left(g \frac{h_1 h_2}{h_1 + h_2} \frac{\rho_2 - \rho_1}{\rho_2} \right)} \qquad (8.15)$$

where h_1 and h_2 are the depths of the top and bottom layer, respectively. The ocean may be considered, albeit very simply, as a two-layer fluid with warm thermocline water with density $\rho_1 \simeq 1025\,\text{kg m}^{-3}$, overlying deep, cold abyssal waters with density $\rho_2 \sim 1027\,\text{kg m}^{-3}$. If $h_1 = 500\,\text{m}$ and $h_2 = 3500\,\text{m}$, then, from equation 8.15, $c = 2.9\,\text{m s}^{-1}$. This wave speed can be compared with a long surface gravity wave, such as a tsunami travelling at $200\,\text{m s}^{-1}$. For a long internal wave having a period of $1000\,\text{s}$, the wavelength will be $2.9\,\text{km}$, which may be compared with a surface wavelength of $200\,\text{km}$.

The amplitude of a typical long internal wave can be estimated in the following way. The hydrostatic pressure variations due to a surface wave of amplitude a_s will be $g\rho a_s$. The corresponding pressure variations due to an internal wave of amplitude a_I are $g[(\rho_2 - \rho_1)/\rho_2)]a_I$. If it is assumed that hydrostatic pressure changes will be similar for both waves, then:

$$a_I = \frac{\rho_2}{\rho_2 - \rho_1} a_s \qquad (8.16)$$

From the previous example, $\rho_2/(\rho_2 - \rho_1)$ is 513, and therefore a long surface wave of height $20\,\text{cm}$ may produce an internal wave approximately $102\,\text{m}$ in height.

In the above examples it has been assumed that the ocean approximates to a two-layer fluid. In reality, the density field changes continuously with depth, and therefore internal waves are rather more complicated than previously suggested. In general, the largest amplitude internal waves are located in regions where the density gradient is a maximum, such as the base of the seasonal thermocline or the halocline in the Arctic Ocean. In regions of continuous density variation, the buoyancy frequency described by equation 2.9 is a useful parameter. The buoyancy frequency corresponds to the highest frequency of the internal waves that can occur. Also, by analogy with equation 2.9, it can be appreciated that the vertical variation in density will produce corresponding variations in the buoyancy frequency.

Hence the internal waves will propagate at difference speeds and refraction of the wave fronts will occur. The internal waves will, therefore, generally propagate both horizontally and vertically. An internal wave generated in the thermocline will tend to propagate downwards into the deeper ocean.

The mechanisms that produce internal waves are numerous. Internal waves in the surface thermocline are often caused by storms, but tides and surface currents may also generate them, particularly close to the shelf break or at the entrance to an estuary. They are found literally everywhere in the ocean and often, because of their large amplitudes, cause problems for physical oceanographers. For example, hydrographic data may be contaminated by internal waves, which may result in considerable errors in the determination of geostrophic currents by the dynamic method. In addition, sound waves may be refracted by a group of internal waves which will add noise to the received signal.

8.5 Ocean tides

The tides are ubiquitous in the ocean. In the deep ocean, tidal currents ranging between 1 and $10 \, \text{cm s}^{-1}$ are observed in most current meter records and in the trajectories of neutrally buoyant floats. To the dynamical oceanographer, these tidal fluctuations represent an annoyance and they have to be removed from the record to obtain long-term currents. Measurements by bottom pressure gauges, and more recently by satellite altimeters, have shown for the first time the regular rise and fall of the tides, typically 10–100 cm in amplitude, over large areas of the deep ocean. Over the continental shelves and especially in shallow, semi-enclosed seas, such as the North Sea, the tidal ranges are an order of magnitude larger than those in the deep ocean. Amplitudes of over 10 m occur in the Severn Estuary, United Kingdom, whilst in the Bay of Fundy, Nova Scotia, the amplitude is 14 m. The associated tidal currents are usually in the range from 1 to $2 \, \text{m s}^{-1}$, and are generally larger than wind-induced currents and geostrophic currents. In the English Channel, between the island of Alderney and Cap de la Hague, tidal currents up to $4.5 \, \text{m s}^{-1}$ occur, whilst near Bodø, north Norway, at the entrance to the Skjerstadt Fjord, currents of $8 \, \text{m s}^{-1}$ are regularly observed.

The relationship between the tides and the Moon's position has been known since antiquity. Early Roman writers described the twice-monthly spring and neap cycle, and a Persian philosopher, in the second century BC, showed that the difference between the height of consecutive tides, known as the diurnal inequality, varied with the Moon's position north and south of the equator. The first theory of tides was proposed by Newton (1687), following Galileo's ideas on orbital motion. Newton recognized that the orbital motion of the Moon around the Earth, though itself in equilibrium, would produce an imbalance of forces on the Earth's surface which would, in turn, distort the sea surface.

Figure 8.9 shows the net forces produced by the Moon at different points on the surface of the Earth. If the Earth and the Moon are considered as point masses, then it can be shown that they will rotate about their common centre of gravity with the outward centrifugal force balanced by the central force of gravitational attraction. However, on a sphere, as opposed to a point mass, this balance of forces is only achieved at the centre of mass of the Earth. At a point Z, the zenith on the Earth's surface, the gravitational attraction is greater than that at the centre of the Earth because of its close proximity to the Moon, whilst at the point N, the nadir, the gravitational attraction is less than at the centre of the Earth. Furthermore, because all points on

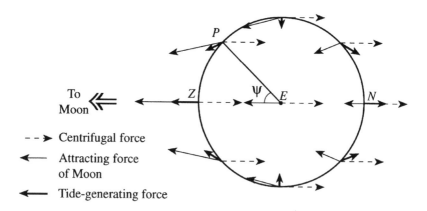

Figure 8.9   Tide-generating force as a resultant of attracting force and centrifugal force along a meridional section through the Earth. Z and N indicate the Moon in zenith or nadir, respectively; E is the centre of the Earth

the Earth describe circles of the same radius as the Earth rotates about the common centre of gravity of the Moon and the Earth, the outward centrifugal force at each point on the Earth's surface is the same. Therefore, at the point Z, the central gravitational force exceeds the centrifugal force and there is a net force directed towards the Moon. At the point N, the centrifugal force exceeds the central gravitational force and there is, therefore, a net force directed away from the Moon.

The net force F per unit mass acting at the point Z is given by the difference between the gravitational force at the centre of the Earth and that at the point Z. The force per unit mass, F_e, at the centre of the Earth is given by:

$$F_e = \frac{Gm_L}{R^2} \qquad (8.17)$$

where G is the gravitational constant, m_L is the mass of the Moon, and R is the distance between the centres of the Moon and the Earth. Since the distance to the point Z is $R - a$, where a is the radius of the Earth, the net force per unit mass at Z is:

$$F = Gm_L \left[\frac{1}{(R-a)^2} - \frac{1}{R^2} \right] \qquad (8.18a)$$

or

$$F = \frac{Gm_L}{R^2} \left[\frac{1}{(1 - a/R)^2} - 1 \right] \qquad (8.18b)$$

Since $a/R = 1/60$, then by the binomial theorem:

$$\frac{1}{(1 - a/R)^2} = 1 + \frac{2a}{R} + \dots$$

Hence, neglecting all higher-order terms:

$$F = \frac{Gm_L}{R^3} 2a \qquad (8.19)$$

It is noted that the net tidal force depends on the inverse cube of the Earth–Moon distance, whilst the gravitational force depends on the inverse square.

For a general point, P, on the Earth's surface, the tidal force can be resolved into vertical (Z_1) and horizontal (H_1) components. It can be shown that these components are given by:

$$Z_1 = \frac{-2Gm_La}{R^3} \frac{3}{4} \left[\cos 2\psi + \frac{1}{3} \right] \qquad (8.20)$$

$$H_1 = \frac{2Gm_La}{R^3} \frac{3}{4} \left[\sin 2\psi \right] \qquad (8.21)$$

It is noted that ψ will only correspond with the latitude when the Moon is overhead at the equator, as depicted in Figure 8.9. Numerical evaluation of Z_1 shows that the component is typically 9×10^{-6} of the Earth's acceleration due to gravity. Therefore, the tidal force in the direction of the local vertical can be neglected compared with g. Only in the horizontal direction is the tidal-force component comparable with other horizontal forces such as the pressure-gradient force. This is an analogous situation to that described in Section 7.1 where the vertical component of the Coriolis force was neglected and the horizontal component retained. The horizontal components of the tidal-generating force, shown in Figure 8.9, are directed towards the zenith and nadir. At Z and N, the horizontal component vanishes since $\psi = 0$. This convergence of forces will cause a horizontal acceleration towards the zenith and nadir, and in consequence, a rise in sea level will occur at these two points. A reduction in sea level will occur along the great circle path at 90º longitude to the zenith and nadir. The Moon's orbit varies relative to the Earth's equator, and so the latitude of the zenith and nadir will change with time, but the relative distribution of tidal forces will remain similar. In addition, it is noted that the rotation of the Earth does not affect the tide-producing forces and therefore, as the Earth rotates, the tide will appear to rotate around the globe with the Moon. The apparent period of rotation of the Moon around the Earth is 24 hours and 50 min, and therefore the semi-diurnal tide will be delayed by 50 min each solar day. Hence, it will be appreciated that consecutive high waters will be delayed by 25 min.

In addition to the semi-diurnal tide, variations in the Moon's position will produce other longer-period tides. For example, the distance between the Earth and the Moon varies with a period of 27.55 days, and a larger range of tides will occur when the Earth is closest to the Moon. In addition, the angle between the position of the Moon and the Earth's equatorial plane, known as the declination angle, has a period of 27.32 days and this variation also affects the tidal ranges.

The tidal forces due to the Sun can be obtained in exactly the same way as described for the Moon. The ratio of the tidal force produced by the Sun compared with that of the Moon is given by $(m_S/m_L)(r_L^3/r_S^3)$ from equation 8.19, where m_S and m_L are the masses of the Sun and Moon, respectively, and r_S and r_L are the respective orbital distances from the Earth. The ratio has a value of 0.46, and therefore the lunar tides will always dominate the solar tides. These solar tides will have semi-diurnal and diurnal periods, as well as a semi-annual period which is associated with annual variations in the Earth–Sun distance and in the declination angle. Table 8.2 shows the principal tidal frequencies associated with the orbital variations of both the Moon and the Sun relative to the Earth. These tidal frequencies are known as the astronomical tides. Detailed astronomical calculations of the orbits of the Sun, Earth and Moon have shown that 396 distinct tidal frequencies can be identified.

Despite the considerable qualitative success of the astronomical theory of tides, the observations of tides, which commenced in the seventeenth century, show considerable discrepancies with the theory. First, the amplitude of the surface elevation for the semi-diurnal, or M_2, tide is $25 \cos^2 \psi$ cm according to the astronomical theory. For $\psi = 45^\circ$, the theoretical amplitude is 12.5 cm, whilst observations of the M_2 tide in the North Atlantic Ocean reveal amplitudes which are an order of magnitude higher. Only in the Mediterranean Sea is the observed tide of the predicted amplitude. Secondly, the M_2 high tide at a

Table 8.2 Principal astronomical tides

Tide	Generating force	Type	Period
M_2	Moon	Semi-diurnal	12 hours 25 minutes
S_2	Sun	Semi-diurnal	12 hours 0 minutes
N_2	Moon	Semi-diurnal	12 hours 40 minutes
K_2	Moon and Sun	Semi-diurnal	11 hours 58 minutes
K_1	Moon and Sun	Diurnal	23 hours 56 minutes
O_1	Moon	Diurnal	25 hours 49 minutes
P_1	Sun	Diurnal	24 hours 4 minutes

particular place should, in theory, correspond to the passage of the Moon over the meridian, but observations show a delay in the time of high water by as much as a few hours. The magnitude of this delay varies from place to place. Thus, the 'equilibrium theory', though important as a qualitative description of tides, could not be used for tidal prediction.

In the eighteenth and nineteenth centuries, it was recognized by Laplace and other mathematicians that the tidal wave would be distorted by:

(i) The horizontal component of the Coriolis force.
(ii) The distribution of the continents.
(iii) The variation in the depth of water as the tide travels around the globe.

Therefore, for a complete theory of tides, it would be necessary to solve the complete set of hydrodynamic equations on a rotating sphere subject to the astronomical tide-generating force. Although the mathematicians Laplace and Hough did find solutions to these equations, they were too simple to describe the complex tidal response of an ocean basin with complicated bottom topography and irregular coastlines. The gulf between tidal theory and observation prompted the adoption of a different approach to tidal predictions which is still widely used today. This approach is known as harmonic analysis. It had long been recognized that, although the response of the ocean to the astronomical tidal forces was complicated, each unique astronomical frequency would produce a corresponding signal in each tidal record, provided that the ocean behaved as a linear system. Thus, a tidal record could be described by a summation of sinusoidal waves having astronomical frequencies (Table 8.2) and, for N frequencies, the sea surface elevation is given by:

$$\xi = \sum_{i=1}^{N} a_1 \cos(\sigma_i t - \phi_i) \tag{8.22}$$

where a_i and ϕ_i are the amplitude and phase corresponding to an astronomical frequency σ_i. In order to calculate a_i and ϕ_i, it is necessary to have records at least 10 tidal periods long and, in practice, because of the spring–neap cycle, it is usual to take a minimum record of 28 days' duration.

Table 8.3 Amplitudes of constituent tides at Immingham, UK, and Manila, Philippines

	M_2	S_2	N_2	K_2	L_2	O_1	K_1
Immingham (amplitude, cm)	223.2	72.8	44.9	18.3	12.1	16.4	14.6
Manila (amplitude, cm)	20.3	–	–	–	–	28.3	29.7

Table 8.3 shows the harmonic analysis of tidal records from Immingham (UK) and Manila (Philippines). It can be seen that, at Immingham, the semi-diurnal, or M_2, tide is dominant, whilst at Manila the diurnal, or K_1, tide is more important. The harmonic analysis shows that, in addition to the dominant components, there are a number of other frequencies present. At Immingham, the S_2, N_2 and K_2 semi-diurnal tides are present, and in all it is possible to distinguish 23 harmonic components. At Manila, another diurnal component, O_1, can be seen, whilst the only semi-diurnal tide is M_2.

Having obtained the phases and amplitudes of the tides for each astronomical frequency, it is then possible to make tidal predictions indefinitely into the future. However, in harmonic analysis, other components appear which do not correspond with astronomical frequencies. First, seasonal variations in wind stress, atmospheric pressure and ocean temperature all cause annual fluctuations in sea level. Secondly, long-term changes in sea level may be present in a tidal record as the result of climatic variations and relative movement of the land. Thirdly, the hypothesis that the ocean is a linear system is not valid in shallow water, i.e. water less than 100 m deep where large-amplitude tidal currents occur. Non-linear interactions between the tidal components produce higher-frequency harmonics. For example, the M_4 and M_6 tides are harmonics of the M_2 semi-diurnal tide and they have periods of 6 hours 12 min and 3 hours 6 min, respectively. In the English Channel, the M_4 tide has an amplitude as large as the M_2 component and it produces a double high water at a number of ports. Indeed, at Southampton, the M_2 and M_4 interaction gives an eight-hour-long stand of high water. Other major harmonics occur between M_2 and S_2, giving frequencies of $M_2 - S_2$ and $2M_2 - S_2$.

From the harmonic analysis, it is a straightforward process to obtain the range of spring and neap tides. For a station such as Immingham, where S_2 and M_2 are the dominant constituents, the range of the spring tide, when the tidal forces of the Sun and the Moon are in phase, is given by $2(a_{M_2} + a_{S_2})$. The range of the neap tide at such a station is $2(a_{M_2} - a_{S_2})$ and, at this time, the Sun and the Moon are 90° out of phase. For Immingham (Table 8.3), the spring range is 5.92 m and the neap range is 3.01 m.

Harmonic amplitudes and phases for a given frequency can be used to obtain a geographical representation of the tide in an ocean basin or a sea. The lines of constant phase are known as co-tidal lines and they correspond to all points which are the same tidal stage relative to the Moon's position with respect to the Greenwich meridian. The lines of constant amplitude, or co-range, correspond to points having the same amplitude. Figure 8.10 shows co-tidal and co-amplitude lines in the Atlantic Ocean, predicted by a model, for the M_2 tide. It can be seen that the M_2 tide rotates in an anticlockwise direction about a point in the centre of the North Atlantic Ocean, known as the amphidromic point. At this point the tidal range is a minimum. Typical amplitudes in the Atlantic Ocean range from 50 cm in the centre up to 150 cm on the eastern Atlantic coast. The highest tidal amplitudes tend to be found close to the continental coasts. Generally, amphidromic points occur in the central parts of ocean basins and this is why mid-ocean islands often have small tidal ranges. In the North Sea, two amphidromic points exist. One is located in the southern North Sea and the other is just off the southern Norwegian coast, and the M_2 tides rotate in an anticlockwise direction around the North Sea. Measurements made using bottom pressure recorders in the deep ocean have verified the existence of amphidromic points in all of the ocean basins.

As briefly discussed earlier, the equilibrium theory of tides cannot account for the amphidromic points and the observed ranges in sea surface elevation. This arises because it is not possible for the sea surface to come into equilibrium with the tide-generating forces on the short time scale of the major tides. An ocean forced by tidal frequencies will produce long gravity waves whose frequencies will be similar to the Coriolis frequency. Such long gravity waves will, therefore, be considerably modified by the horizontal component of the Coriolis force. In a semi-enclosed basin these waves will propagate as Kelvin waves (described in Section 8.3), and they will be reflected and rotated by the boundary. This rotation will be in an anticlockwise direction in

Figure 8.10   Co-tidal map of the M_2 (semi-diurnal lunar) tide for the Atlantic Ocean. Solid lines: co-tidal lines (phase) in degrees. Dashed lines: co-amplitude lines in centimetres. The direction of rotation of the tidal wave is shown by a solid arrow

the Northern Hemisphere and this corresponds, in a qualitative manner, to the behaviour of the M_2 tide in the North Atlantic Ocean.

During the nineteenth and twentieth centuries, many mathematicians studied the behaviour of long waves in a variety of idealized basins of uniform depth by analytical methods. More recently, however, direct numerical solutions of the linear hydrodynamic equations for the world's oceans, including realistic bottom topography and coastlines, have been obtained by forcing the models with astronomical tidal forces. These models have shown that good agreement with observations can be obtained, as had originally been suggested by Laplace (1778). However, for the tides on the continental shelves, there exist non-linear interactions. The variations in bottom topography are also important in these regions and frictional processes play a crucial role. For all the above reasons, more complicated models are required to give reasonable predictions of tidal elevations and currents on the continental shelves.

8.6 Storm surges

Tidal theory can give predictions of sea level to an accuracy of a few centimetres. However, systematic departures from tidal theory predictions do occur as the result of variations in surface pressure and wind stress. Generally, these departures are of the order of a few centimetres and they go largely unnoticed. Occasionally, however, they may be of the order of 1 m and, in combination with a high spring tide, can produce considerable flooding of low-lying coastal regions. On 31 January and 1 February 1953, one such storm surge in the North Sea reached a height of between 2 and 4 m above the predicted sea level, and inundated large areas of eastern England and the Netherlands, with a loss of 1700 lives. Surges also occur in other parts of the world where there are partially enclosed seas which are relatively shallow. The northern Adriatic Sea, including the Venetian region, is vulnerable to surges caused by intense winter depressions in the western Mediterranean Sea. Tropical cyclones in the Bay of Bengal have been known to produce surges of between 4 and 7 m along the coast of Bangladesh. These latter surges are particularly devastating because of the low relief of the Ganges delta and they have been known to flood areas up to 100 km inland. Although positive surges in sea level are of importance to those on land, negative surges can be equally important to the master of a ship. For instance, a negative surge of 1–2 m will cause difficulties for a large oil tanker navigating a shallow entrance to a port and, in exceptional cases, it may lead to a ship floundering on a sandbank.

A 100 Pa reduction in surface pressure will cause a 1 cm rise in sea level, assuming that the water is in hydrostatic equilibrium. An intense depression, similar to that associated with the 1953 storm surge in the North Sea (Figure 8.11), having a central pressure of 97 kPa will produce a rise in sea level of approximately 42 cm, assuming a mean surface pressure of 101.2 kPa. This rise in sea level is about one order of magnitude less than the observed value, and it is therefore apparent that most of the surge must be accounted for by the direct effect of wind stress on the sea surface. Consider now a steady wind stress, τ, blowing against a coast. In equilibrium, the wind-stress force must balance the pressure gradient caused by the slope in the sea surface, $\partial \xi / \partial x$:

$$ g \frac{\partial \xi}{\partial x} = \frac{\tau}{\rho h} \qquad (8.23) $$

where h is the depth of the sea. For a given wind stress, the slope in the sea level is inversely proportional to the depth of the sea. Thus, it is only for shallow seas that a major change in sea level is expected.

A steady wind of $22\,\mathrm{m\,s^{-1}}$, or 43 knots, will produce a stress of approximately $0.9\,\mathrm{N\,m^{-2}}$ using the bulk aerodynamic formula (equation 5.9). For the North Sea, $h \sim 40\,\mathrm{m}$ and therefore, from equation 8.23, the sea surface slope is 2.2×10^{-6}. The North Sea is about 600 km in length, and therefore the increased height of the sea surface in the southern North Sea, produced by the southward wind stress, would be approximately 1.3 m. Hence, it can be seen that a surface-wind stress similar to that which produced the 1953 storm surge can account for a much higher sea surface elevation than pressure factors alone. The total estimated rise in sea level due to both wind stress and pressure is approximately 1.7 m, which is lower than that observed along the English and Dutch coasts in 1953. The reason for this discrepancy is related to the effect of the focusing of the surge by estuaries, by the very shallow depth of water along the southern North Sea and by time-dependent dynamics.

An additional feature of North Sea surges is that they propagate in an anticlockwise direction around the coasts. Figure 8.12 shows the surge elevation on 31 January and 1 February 1953 for the North Sea. The surge was first observed at Aberdeen at 14.00 hours on 31 January 1953, with a height of 0.6 m, and subsequently travelled down the east

Figure 8.11   The North Sea floods of 1953. Meteorological charts for the period 12.00 hours GMT on 30 January to 00.00 hours GMT on 1 February 1953 (contour interval 4 mb)

Figure 8.12   The North Sea floods of 1953. Surge elevations computed from a numerical model for the period 12.00 hours GMT on 31 January to 00.00 hours GMT on 2 February (contour interval 25 cm)

1200 31/1/53

1800 31/1/53

0000 1/2/53

0600 1/2/53

1200 1/2/53

0000 2/2/53

coast and arrived at Chatham 11 hours later, when its height was 2.5 m. It then followed the Belgian and Dutch coasts where it reached a maximum height of 3.2 m. Its propagation was very similar to that of the semi-diurnal tide, which also travels anticlockwise about the North Sea. This surge propagation is related to the Coriolis force, which has been neglected in equation 8.23. First, the Coriolis force causes a rotation of the wind-driven current to the right in the Northern Hemisphere, and thus a southward wind stress induces a transport of water towards the eastern coast of Britain which, in turn, induces a rise in sea level. This rise in sea level is then transmitted along the coast by a Kelvin wave, which is trapped to the coast by the Coriolis force. The speed of the Kelvin wave for the North Sea, assuming a depth of 40 m, is calculated to be $20\,\mathrm{m\,s}^{-1}$ (in an anticlockwise direction). The coastal distance between Aberdeen and Chatham is 800 km, and therefore the expected time delay, for the theoretical speed of $20\,\mathrm{m\,s}^{-1}$, is 11.1 hours, which is remarkably close to the observed value. Thus, it has been shown that wind stress, surface pressure and the Coriolis force can account for many features of North Sea storm surges.

The southward Kelvin wave propagation makes it possible to predict storm surges on the east coast of England some 6–12 hours in advance from the observation of tidal residuals (i.e. the difference between the actual sea level and the sea level predicted from tide tables) on the north-east coast of Scotland. This technique has been used routinely for sea-level prediction by the Meteorological Office during recent decades. However, numerical models using the 24-hour forecasted winds are now being used to obtain accurate predictions for all parts of the North Sea. Similar techniques have also been used in the Adriatic Sea to give predictions of sea level for up to 12 hours ahead. All forecasts are limited by the predictability of the position and intensity of the storm in question. For regions such as the Bay of Bengal, accurate sea-level predictions will require detailed forecasts of the paths of tropical cyclones.

9

Energy Transfer in the Ocean–Atmosphere System

9.1 Modes of energy in the ocean–atmosphere system

Table 9.1 shows the magnitude of the major energy sources and sinks in the ocean–atmosphere system. It shows that the major fraction of the incident solar energy is used in the evaporation of water into the atmosphere whilst the kinetic energy of the winds, dissipated into heat by friction, is a relatively minor fraction of the solar flux. However, even the dissipation of the winds' kinetic energy is at least one order of magnitude larger than the other energy sources listed. The direct conversion of visible solar radiation into carbohydrate by photosynthesis is seen to be a relatively inefficient process when compared with the total flux of radiation. Its estimated value, which is based on values of photosynthesis on land, is probably too large because primary productivity in the oceans is limited by a number of

Table 9.1 Magnitude of major energy sources and sinks in the Earth–ocean–atmosphere system

	Energy flux ($W\,m^{-2}$)
Solar radiation	340
Latent heat	70
Rate of kinetic energy dissipation	~2
Photosynthesis	~0.1
Geothermal heat flux	0.06
World energy production	0.02
Solar reflection from full Moon	0.014
Ocean tides	~0.006

factors, such as the supply of nutrients. Most of the energy stored by photosynthesis will be returned to the ocean or atmosphere by oxidation of organic matter, either by burning or by decomposition, and only a small fraction will be laid down in sediments as potential fossil fuels. It is interesting to note that world energy production by the burning of fossil fuels, though small, is now a non-negligible fraction of photosynthetic production. It can be seen from Table 9.1 that the world energy production would, however, have to increase one-hundredfold before it matched the dissipation of kinetic energy by the winds. Geothermal heat, produced by radioactive decay in the Earth's interior, is very small on the world-wide scale, even though it can be locally large in volcanic regions. Dissipation of energy by ocean tides is virtually insignificant on a global scale. However, the majority of the dissipation occurs on the continental shelf, especially in the Bering Sea, the Sea of Okhotsk, the Irish Sea and the Patagonian Shelf where energy fluxes are of the order of 10^3 times larger than the world-wide value.

In the discussion of the energy budget of the ocean and atmosphere, it is therefore possible to neglect all of the sources of heat except the incident flux of solar energy. Reference to Figure 3.10 will show that, apart from the 30% of the incident solar flux which is reflected back into space, the majority of the incident solar radiation is absorbed by the atmosphere and the ocean. However, only 15% of the solar radiation is directly absorbed by the atmosphere, and the remaining energy is converted into latent heat of condensation and fusion, turbulent heat conduction and long-wave radiation at the surface, to be subsequently released or absorbed into the atmosphere. From the first law of thermodynamics, the energy absorbed can reside as either internal (heat) energy or it can be used to do work against the environment, appearing as potential or kinetic energy. Consider a volume of fluid which is heated uniformly. The temperature, and therefore the internal energy, of the fluid will increase. At the same time, the fluid will expand and do work against gravity. The potential energy of the fluid will therefore increase. The internal energy per unit mass (IE) is $C_v T$, where C_v is the specific heat at constant volume and T is the absolute temperature. The potential energy per unit mass (PE) is gz, where g is the acceleration due to gravity and z is the height above sea level. For a column of fluid of unit cross-section, the mass is $\rho\,dz$, where ρ is the fluid density, and therefore:

$$\text{IE} = \int_0^h C_v T \rho \; dz \tag{9.1}$$

and

$$\text{PE} = \int_0^h g z \rho \; dz \tag{9.2}$$

where h is the depth of the fluid.

From the hydrostatic equation, $p = \rho g z$ and, for the atmosphere, $p = \rho R T$, and thus:

$$\text{PE} = \int_0^h R T \rho \; dz \tag{9.3}$$

The ratio of the potential energy to the internal energy for the atmosphere is therefore R/C_v or $\frac{2}{5}$. Thus, for each unit of heat absorbed in the atmosphere, $\frac{5}{7}$ of the heat will go into internal energy and $\frac{2}{7}$ will go into potential energy. If it is assumed that 70% of the incident solar radiation is absorbed both directly and indirectly by the atmosphere, then the fraction of solar radiation going into potential energy is $0.7 \times \frac{2}{7}$ or 0.2 of the incident solar flux.

The ocean has a small compressibility, and therefore virtually all of the absorbed solar radiation will appear as internal energy and only 0.01%, a negligible fraction, will go into potential energy. This potential energy can be measured by sea-level variations. For example, the seasonal cycle of temperature can produce changes in the sea level of a few centimetres.

Returning once more to the atmosphere, it is noted that, for the wind circulation to be in a steady-state balance, the energy dissipated by the wind must be balanced by an equal energy input which must, in turn, be derived from the potential energy of the atmosphere. The conversion of PE into kinetic energy (KE) is about $2\,\text{W}\,\text{m}^{-2}$, from Table 9.1, and this may be compared with a rate of $68\,\text{W}\,\text{m}^{-2}$ for the production of PE. Therefore, only 3% of the potential energy is used to drive the general circulation of the atmosphere. For the potential energy and internal energy to be in a steady-state balance, the majority of the potential and internal energy must be continuously and directly dissipated by long-wave radiation into space without being involved in the generation of kinetic energy. In thermodynamic terms, the atmosphere heat engine has a very low efficiency.

To understand why the atmosphere is so wasteful of its potential energy, consider two immiscible fluids of different density, ρ_1 and ρ_2, as shown in Figure 9.1a. The fluids are separated by a vertical partition and have equal volume. It can be shown that the potential energy of the system is $(gh^2/2)(\rho_1 + \rho_2)$, where h is the depth of the fluid. If the partition is now removed, the denser fluid will flow under the lighter fluid and it will eventually take up the position shown in Figure 9.1b. During this process, kinetic energy will be generated and it will, in its turn, be dissipated by the internal viscosities of the two fluids. In its final state the system still has potential energy but, because the interface is parallel to the geopotential surface, it is not possible to derive any more kinetic energy from the potential energy field. This remaining potential energy is known as unavailable potential energy and it can be shown to be equal to $(gh^2/4)(\rho_2 + 3\rho_1)$. The difference between the initial and final potential energy is $(gh^2/4)(\rho_2 - \rho_1)$. It is this fraction of potential energy which is converted into kinetic energy and it is known as the available potential energy (APE). The ratio of kinetic energy to potential energy is $\frac{1}{2}[(\rho_2 - \rho_1)/(\rho_2 + \rho_1)]$. If the density difference between the two fluids is small, then the APE will be

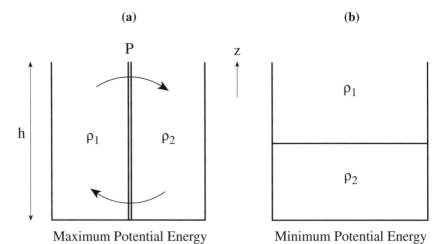

(a)   (b)

Maximum Potential Energy   Minimum Potential Energy

Figure 9.1   Potential energy in a two-layer fluid. (a) Maximum potential energy, before partition removed. (b) Minimum potential energy much less than the initial potential energy. In the atmosphere, the horizontal variations in air density are relatively small when compared with the mean density, being typically less than 5%. Therefore, the ratio of kinetic energy to potential energy is approximately 1%. In the ocean, where the above model is a better analogue, density variations are less than 0.5%, and therefore the ratio of kinetic energy to potential energy is about 0.1%.

The APE is therefore proportional to the horizontal variations in density in both the atmosphere and the ocean. In the atmosphere, the most satisfactory indicator of APE is the potential temperature because it avoids the adiabatic effect (Figure 9.2). The horizontal gradient of potential temperature, and therefore APE, is caused by the difference in net radiation between the equator and the poles. The largest slope is in the middle latitudes. Most of the equatorial heating occurs at low levels, whilst the polar cooling occurs at high levels in the troposphere. The potential temperature gradient is therefore between the lower equatorial troposphere and the upper polar troposphere. In the ocean, the appropriate indicator of APE is the potential density, although the potential temperature generally shows the main features of the distributions. The majority of the APE is associated with the slope of the main thermocline which slopes upwards both polewards and equatorwards from the centres of the sub-tropical gyres. In the Atlantic

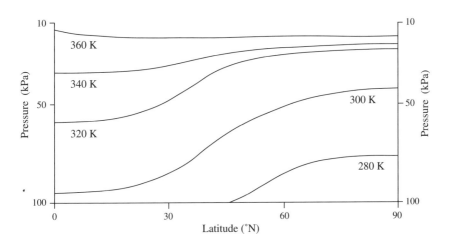

Figure 9.2 Longitudinally averaged atmospheric potential temperature

and Pacific Oceans, the largest APE is concentrated on the poleward boundaries of the gyres where the main thermocline rises from its deepest point to the surface in a horizontal distance of 200–300 km. Large slopes in the density surface also occur across the sub-tropical and Antarctic convergence zones in the Antarctic Circumpolar Current. In deep water the density surfaces are close to horizontal and therefore, in common with the above analogue, there is little APE in the abyssal layers.

Figure 9.3 shows the energy cycle of the atmosphere and ocean in terms of the KE and APE reservoirs. For the atmosphere, the APE is between three and four times larger than the KE and the total energy of the system is approximately $700 \times 10^4 \, \mathrm{J \, m^{-2}}$. If the energy of the atmosphere was not replenished by radiation, then all of its energy would be dissipated on a time scale of $(700 \times 10^4)/2.3\mathrm{s} \simeq 35$ days, which is very similar to the radiative time scale discussed in Chapter 2.

The APE and KE reservoirs of the ocean are not as certain as those given for the atmosphere, and therefore Figure 9.3 shows only some probable estimates obtained from a variety of investigations. The KE estimate is based on typical velocities of $10 \, \mathrm{cm \, s^{-1}}$ for the interior ocean and the APE is based on a model of the world ocean. It is noted that the ratio of APE to KE is approximately 25 : 1, and therefore most of the energy of the ocean is stored as APE. This APE may be regarded as the flywheel of the ocean. However, the total energy reservoir of the ocean is less than 10% of the energy of the atmosphere, and so the atmosphere is the energetically dominant system. There are two sources of energy for the ocean. One source is the direct driving of ocean currents by the wind stress. The other source is through the surface buoyancy flux which, in part, drives the thermohaline circulation. A surface wind of $10 \, \mathrm{m \, s^{-1}}$, typical of middle latitudes, will cause an energy transfer of approximately $1 \, \mathrm{W \, m^{-2}}$ into the oceans. Of this energy input, some will be dissipated quickly in the surface layer, a fraction will go into the surface waves and swell, some will go into internal waves and the remainder will go into ocean currents. The energy transfer into ocean currents is probably about 0.3% of the work done by the wind stress. The energy input by the surface buoyancy flux is not certain. One study has shown that the APE generated by surface cooling in the Northern Hemisphere is less than the energy input by the surface winds. Therefore, the ocean is rather different from the atmosphere because the majority of its APE is

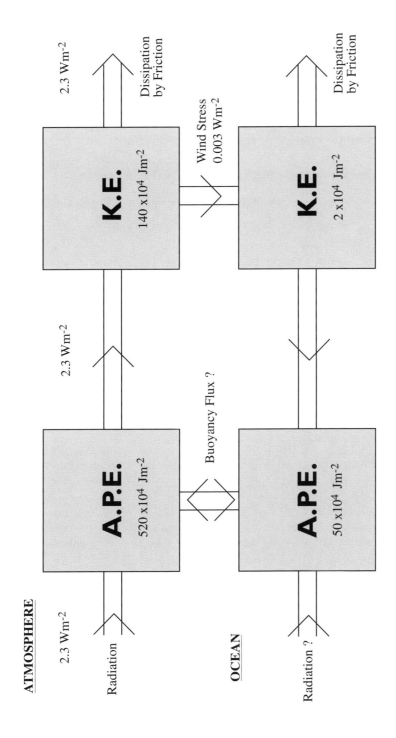

Figure 9.3 Energy diagram of the atmosphere and ocean. Note that the estimates of energy in the ocean are uncertain

derived from the kinetic energy input by the winds. Assuming that the dissipation rate is $0.3 \times 10^{-2}\,\mathrm{W\,m^{-2}}$, then the ocean would lose all of its APE and KE in about five years if it were not being continually replenished by the surface wind stress and surface buoyancy flux.

9.2   The kinetic energy of the atmosphere and ocean

The kinetic energy can be partitioned into two components:

(i) The mean kinetic energy (MKE), which is associated with the long-term average circulation of the ocean or atmosphere, thus:

$$\text{MKE per unit volume} = \frac{1}{2}\rho\bar{v}^2$$

where $\bar{v}$ is the mean velocity.

(ii) The fluctuating or eddy kinetic energy (EKE), which is associated with the time variability of the currents and the winds, thus:

$$\text{EKE per unit volume} = \frac{1}{2}\rho\overline{v'^2}$$

where v' is the fluctuation velocity.

The actual velocity at any given time, $v(t)$, is therefore given by $v(t) = \bar{v} + v'$.

In the atmosphere the MKE is approximately the same magnitude as the EKE, whilst in the ocean the EKE is about 5–20 times larger than the MKE. For the observed kinetic energy in the atmosphere, the typical mean velocity and fluctuation velocity would be $12\,\mathrm{m\,s^{-1}}$. In the ocean, the mean velocity is between 2 and $4\,\mathrm{cm\,s^{-1}}$ and the corresponding eddy velocity is $10\,\mathrm{cm\,s^{-1}}$.

In both the atmosphere and the ocean there is a high correlation between the magnitude of the MKE and the EKE. This indicates that energy is being continually exchanged between the two modes. In the atmosphere, the majority of the KE is associated with the locations of the jet streams in the upper troposphere between 30 and 50° latitude. The largest MKE occurs in the sub-tropical jet streams, whilst the transient polar jet streams contribute most to the EKE. The equivalent high-energy regions for the ocean are the western boundary currents and the equatorial currents but, in all of these regions, the fluctuating energy is four to five times larger than the MKE. This shows that even the most well-defined 'climatological' currents, such as the Gulf Stream, are regions of considerable variability. In the interior ocean, away from the boundary currents, most of the kinetic energy is in eddies and only a small fraction, usually less than 10%, is found in the mean flow. Generally, both MKE and EKE decrease from the western to the eastern basins of both the Pacific and Atlantic Oceans. In the Antarctic Circumpolar Current, eddies are found along the entire length of the current, and they are most intense near to the water-mass boundaries of the subtropical and Antarctic convergence zones.

It is interesting to compare the relative contributions of the different scales of atmospheric and ocean circulation systems to the overall budget of kinetic energy. Figure 9.4 shows the kinetic energy of a variety of atmospheric phenomena and Figure 9.5 is a similar diagram for the ocean. It can be seen that the most intense atmospheric systems, such as the tornado and the tropical cyclone, are not the most energetic ones. An extra-tropical cyclone has a kinetic energy which is between 10 and 100 times larger than that of a tropical cyclone. The former system has a diameter about 10 times as great as a tropical cyclone and so its larger kinetic energy is readily appreciated. As a further example, a sea-breeze system, which may penetrate inland for up to 100 km, has a kinetic energy equivalent to about 10 000 tornados. Thus, most of the fluctuation energy in the atmosphere is bound up in jet streams, Rossby waves and extra-tropical cyclones, whilst the other scales of motion, although of local importance, make little contribution to the total kinetic energy budget. In the ocean, the scale dependence of the circulation systems' contributions is again apparent. The Antarctic Circumpolar Current, the western boundary currents, such as the Gulf Stream and the Kuroshio, the equatorial undercurrents and the tides are the principal contributors, whilst an individual mesoscale eddy, though locally energetic, has an energy which is only about 5×10^{-5} of the total kinetic energy. Thus, 20 000 mesoscale eddies would be required in the ocean at any one time if their contribution was to be equivalent to the large-scale systems. Ocean eddies have a typical scale of 100 km, and so the requirement would imply that virtually the whole ocean

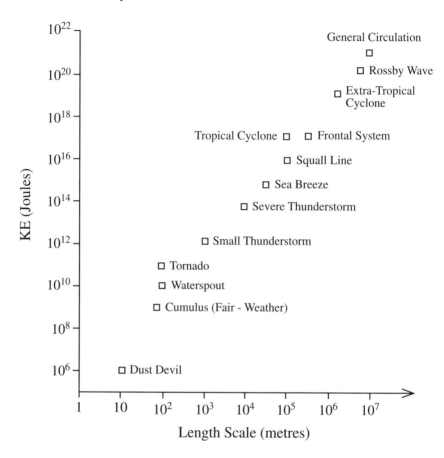

Figure 9.4   Kinetic energy of atmospheric phenomena

would have to be filled with eddies. As yet it is not certain if this is the case, but mesoscale circulations have been found in all of the ocean basins. Less energetic motions are associated with mid-ocean fronts and deep convection which occur on scales between 1 and 10 km. Surface waves and internal wave energy cannot be ignored in the kinetic energy budget. A wind of $10\,\mathrm{m\,s^{-1}}$ will generate an equivalent surface-wave energy of approximately $2 \times 10^4\,\mathrm{J\,m^{-2}}$. If this energy was typical of the whole ocean, it would be of a similar magnitude to the total kinetic energy of the circulation. It is also noted that internal waves have been shown to have a possible maximum energy of $\sim 10^3\,\mathrm{J\,m^{-2}}$, and therefore they may also make a significant contribution to the total energy of the ocean circulation.

Figure 9.5   Kinetic energy of oceanic phenomena

9.3   Mechanisms of kinetic-energy transfer

The simplest method by which kinetic energy is generated from available potential energy is by the mechanism in the tank experiment described in Section 9.1. In this experiment the dense fluid runs under the lighter fluid when the barrier is removed, thus releasing kinetic energy. The vertical circulation that is set up in the tank is known as *direct circulation* because it always converts APE into KE. An example of direct circulation is the Hadley cell, shown in Figure 6.12, where potentially warm air, driven by latent heat release, ascends into the ITCZ. The upward branch moves air polewards in the upper troposphere where it cools by long-wave radiation emission into space, and then it descends in the sub-tropical latitudes. The buoyancy source in the lower equatorial troposphere, i.e. the release of latent heat by condensing water vapour, maintains the circulation, and thus APE is converted into KE. A similar direct circulation can be visualized in the ocean where the sinking of dense water formed as the result of cooling or evaporation occurs, as in the Mediterranean Sea and the Red Sea (Figure 5.9). Therefore, direct circulations are associated with the upward motion of less-dense fluid and the downward motion of denser fluid.

Vertical circulations in which the denser fluid rises and the lighter fluid sinks are known as *indirect circulations*. In such cases, the system increases the APE at the expense of the KE. The Ferrel cell, in the middle latitudes of the troposphere (Figure 6.12), is an example of an indirect, meridional circulation and acts as a sink to the KE of the atmosphere. The loss of KE due to the Ferrel cell is, however, small when compared with the KE gained from the Hadley circulation.

Indirect circulations are responsible for the generation of the large reservoir of APE in the ocean. In Section 7.5 it was shown that Ekman currents, produced by the large-scale wind circulation, cause surface convergence in the sub-tropical zones, and divergence on the equator and in higher latitudes. The vertical Ekman circulation pushes warmer water downwards in the sub-tropical gyre and brings cold water up to the surface in the equatorial zone. It thus increases the APE. As has been shown, a surface-wind stress along the equator will produce a longitudinal tilt in the thermocline in both the Atlantic (Figure 7.18b) and Pacific Oceans, and this will also produce APE. In equatorial regions, because of the direct response of the thermocline to the wind, the APE reservoir can be replenished relatively quickly within a few months. In higher latitudes the process is rather less efficient and it would take in the region of a couple of decades to produce the observed tilt of the permanent thermocline.

The direct circulation is the simplest mechanism by which kinetic energy can be generated from available potential energy. In the atmosphere or in the ocean, there is an additional constraint to the release of PE and that is the rotation of the Earth. Consider what would happen if the tank in Figure 9.1 was rotating when the partition separating the two different density fluids was released. Initially, the behaviour would be similar to the non-rotating case but, after a short time, the Coriolis force would deflect the parcels of fluid perpendicular to the density gradient, as shown in Figure 9.6. In the Northern Hemisphere the dense fluid would be deflected in a westward direction

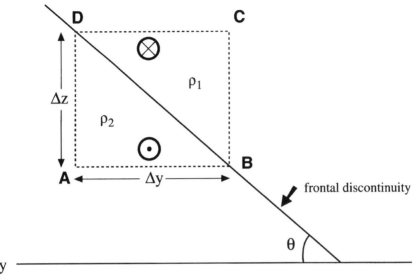

Figure 9.6 Geostrophic equilibrium for a two-layer fluid; y is directed northwards, cross denotes eastward and dot denotes westward geostrophic winds. Note rotation inhibits the adjustment of the fluid to a state of minimum potential energy (see Figure 9.1)

and the lighter fluid would be deflected in an eastward direction. Provided that the tank was an annulus, so that the motion was unimpeded perpendicular to the density gradient, then an equilibrium between the Coriolis force and the pressure gradient would eventually be achieved. The zonal flows would be geostrophic and the interface would slope as shown in the diagram. The rotation of the Earth therefore inhibits the conversion of APE into KE. The slope of the interface can be determined from the geostrophic and hydrostatic equations in the following way. If the geostrophic flow of the dense fluid is u_2 and that of the lighter fluid is u_1, then:

$$f\rho_1 u_1 = \frac{p_C - p_D}{\Delta y} \qquad (9.4a)$$

and:

$$f\rho_2 u_2 = \frac{p_B - p_A}{\Delta y} \qquad (9.4b)$$

using the notation of Figure 9.6. From the hydrostatic equation:

$$p_B = p_C + g\rho_1 \Delta z \tag{9.5a}$$

and

$$p_A = p_D + g\rho_2 \Delta z \tag{9.5b}$$

Subtracting 9.5b from 9.5a and dividing through by Δy:

$$\frac{p_B - p_A}{\Delta y} = \frac{p_C - p_D}{\Delta y} - g(\rho_2 - \rho_1)\frac{\Delta z}{\Delta y} \tag{9.6}$$

Substituting for the geostrophic equation (9.4a and b) into 9.6 to obtain the slope of the interface:

$$\frac{\Delta z}{\Delta y} = \frac{f(\rho_1 u_1 - \rho_2 u_2)}{g(\rho_2 - \rho_1)} \tag{9.7}$$

As density variations are relatively small compared with velocity variations:

$$\rho_1 u_1 - \rho_2 u_2 \simeq \bar{\rho}(u_1 - u_2)$$

where $\bar{\rho} = (\rho_1 + \rho_2)/2$, and hence:

$$\frac{\Delta z}{\Delta y} = \frac{\bar{\rho}f(u_1 - u_2)}{g(\rho_2 - \rho_1)} \tag{9.8}$$

It can be seen that the interface, or front, will have a small slope when the density difference $(\rho_2 - \rho_1)$ is large and the velocity difference $(u_1 - u_2)$ is small. The interface will have a steep slope if the converse is true.

Consider the frontal interface between the northern boundary of the Gulf Stream at approximately 20°C and the Labrador Sea water at about 5°C. Assuming that the Gulf Stream water has an eastward velocity of $0.5 \, \mathrm{m \, s^{-1}}$ and that the cold water is quiescent, then from the equation of state:

$$\rho_1 = 1024 \, \mathrm{kg \, m^{-3}}$$

$$\rho_2 = 1021 \, \mathrm{kg \, m^{-3}}$$

assuming that the temperature dominates the density field. Hence, from equation 9.8:

$$\frac{\Delta z}{\Delta y} = \frac{1022.5 \times 10^{-4} \times 0.5}{9.8 \times 3} \simeq 1.7 \times 10^{-3}$$

Thus, the frontal surface will slope upwards towards the north by 170 m in 100 km.

For the atmosphere, equation 9.8 has to be modified by substitution of the ideal gas equation ($p_1/RT_1 = \rho_1$ and $p_2/RT_2 = \rho_2$) and, with the assumption that at the interface $\rho_1 = \rho_2$, the following expression is obtained:

$$\frac{\Delta z}{\Delta y} = \frac{f\bar{T}}{g}\left(\frac{u_1 - u_2}{T_2 - T_1}\right) \qquad (9.9)$$

where $\bar{T} = (T_1 + T_2)/2$. For the polar front $u_1 - u_2 = 30\,\mathrm{m\,s^{-1}}$, $T_2 - T_1 \sim 15\,\mathrm{K}$, $\bar{T} = 280\,\mathrm{K}$ and

$$\frac{\Delta z}{\Delta y} = \frac{10^{-4} \times 280 \times 30}{9.8 \times 15} \simeq 5.7 \times 10^{-3} \quad \text{or} \quad \frac{1}{175}$$

Therefore, the frontal surface will have a calculated slope of 5.7 km in a horizontal distance of 1000 km, which is similar to the observed value (Figure 9.7).

It has been shown that the effect of the Earth's rotation is to produce regions of discontinuity of density in the ocean and of temperature in the atmosphere. These frontal regions also demarcate the boundaries of different air masses and different water masses. However, these frontal regions are rarely in equilibrium, as described by equations 9.8 and 9.9. Such a balance of forces would not allow for the transfer of heat and other properties across the frontal surface. As an extreme example, if the atmospheric transfer of heat across the 50° latitude circle was stopped, then the atmosphere polewards of 50°N would cool by 100 K in 100 days. This increase in the horizontal gradient of temperature would in turn cause an increase in the upper-level winds relative to the low-level winds and, at a certain point, the wind shear between the upper and lower troposphere would reach a critical value and the flow would become hydrodynamically unstable. This is known as 'baroclinic instability', and Figure 9.8a shows the initial deformation of a frontal

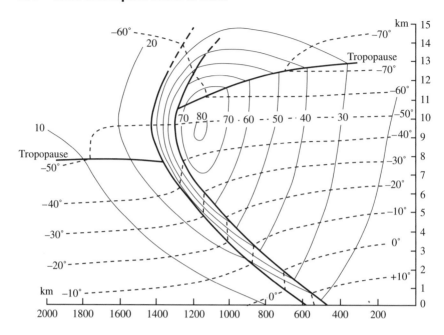

Figure 9.7   Schematic isotherms (dashed lines, °C) and isotachs (thin solid lines, m s⁻¹) in the polar front zone. Heavy lines are tropopauses and boundaries of frontal layer. Note the wind maximum (jet stream) occurs close to the tropopause and on the warm side of the polar front

surface as the result of this instability. The warm air moves polewards and upwards, overriding the cold air, and the cold air moves equatorwards and downwards, undercutting the warm air. If the warm air follows the frontal surface, as shown in Figure 9.8b, an external supply of energy will be required to lift the warm air above the cold air. However, if the warm air follows trajectory AA', it will find itself in a colder environment, and it will accelerate polewards and upwards without any energy input. Similarly, a parcel of cold air following path $A'A$ will find itself in a warmer environment, and it will accelerate equatorwards and downwards. This is very similar to the behaviour of air parcels in a vertically unstable environment where the air parcels continue to accelerate away from their initial positions (when once displaced). All of the parcels in the present example which have a trajectory between the horizontal and the frontal surface will gain kinetic energy from the APE of the front. This kinetic energy will produce a slantwise circulation across the isotherms. Figure 9.9 shows a more detailed picture of the trajectories in an extra-tropical cyclone.

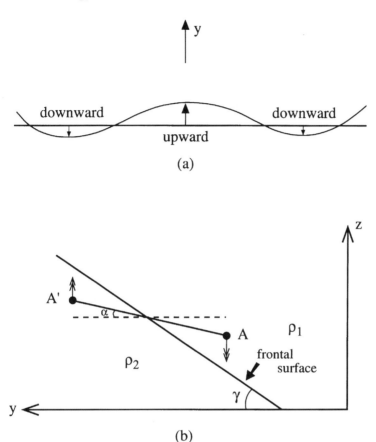

Figure 9.8 (a) Deformation of a frontal surface – plan. (b) Unstable perturbations across a frontal discontinuity. Parcels of air following trajectories AA', will be unstable with respect to their environment, which will cause a conversion of potential energy (PE) to kinetic energy (KE). This will occur if $\alpha < \gamma$

The warm tongue can be seen to move upwards and polewards whilst polar air, developing along the polar front in the rear of the system, can be seen to descend from high levels towards the surface, thus releasing potential energy for the development of the extra-tropical cyclone. Latent-heat release in the region of ascent will provide an additional source of kinetic energy for the system. Eventually, the tongue of warm air will be completely undercut by polar air and the system is then said to be occluded. At this stage most of the APE has been

Figure 9.9 Perspective view of a 100 kPa cyclone and 60 kPa contour pattern. Heavier arrows indicate three-dimensional trajectories in the main ascending and descending branches; thin arrows, their projection on to 100 kPa or 60 kPa surfaces

converted to kinetic energy and the extra-tropical cyclone will begin to decay as surface friction dissipates its kinetic energy. A typical extra-tropical cyclone will grow to its maximum intensity in one to three days and will decay in about five to six days. In some cases the development can be rapid – of the order of a few hours – and this makes accurate prediction of these systems very difficult.

Baroclinic instability is thus occurring at all times in the middle latitudes, and it acts to transfer heat between low and high latitudes. However, this process does show a strong seasonal behaviour as well. In summer, with relatively weak horizontal gradients of temperature, the APE in the atmosphere is smaller and the extra-tropical cyclones are generally weaker. By contrast, in winter, larger horizontal temperature gradients are produced not only between the pole and the equator, but also between the continents and the ocean. These large sources of APE produce intense cyclonic systems which transfer heat both polewards and towards the cold continental land masses. The major baroclinic zones in the Northern Hemisphere are located on the eastern seaboards of the North American and Asian continents in winter, where very cold continental air occurs adjacent to the very warm Gulf Stream and Kuroshio Currents. In conclusion, it is seen that vertical convection in low latitudes, and slant convection in middle and high latitudes, are the major processes by which energy is transferred in the atmosphere.

In the ocean, baroclinic instability is also responsible for the transfer of heat and other properties, such as salinity and nutrients, across frontal surfaces. Figure 9.10a shows the development of a Gulf Stream meander, and the eventual formation of isolated eddies of cold and warm water on either side of the front. The growth rates of these meanders are slower than their atmospheric counterparts, and they take between one week and one month to produce Gulf Stream rings. It is also noted that not all the meanders grow into eddies. The 'cold' Gulf Stream rings move southwards and westwards into the Sargasso Sea, and they may be tracked for up to two years before they finally dissipate. In contrast, the 'warm' Gulf Stream rings, situated on the northern side of the Gulf Stream, tend either to be dissipated on the continental slope by friction or to be reabsorbed into the Gulf Stream system. Mesoscale ocean eddies are also important for the transfer of heat in the higher latitudes of the Southern Hemisphere. The Antarctic Circumpolar Current is the only zonal current which flows unimpeded

(a)

(b)

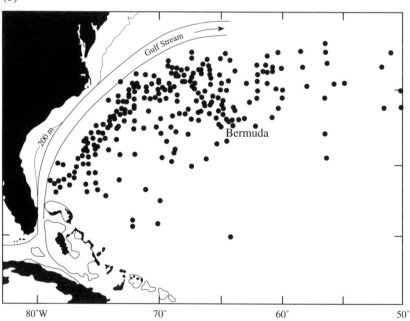

Figure 9.10   (a) The formation of a cyclonic cold ring in the Gulf Stream. (b) The distribution of Gulf Stream cold rings from 1932 to 1976

around the globe, and it therefore has some similarities to the zonal atmospheric flows. Eddies are an important mechanism for transferring heat and salt across the zonal frontal boundaries. The energy for the eddies comes from the slope of the isopycnal (equi-density) surfaces between 40 and 60°S. Recent studies have shown that these Southern Hemisphere eddies may account for a horizontal heat transfer of 0.4×10^{15} W.

9.4   The general circulation of the atmosphere

The atmosphere and the ocean are both highly complex systems which fluctuate on a large variety of time and space scales. Both systems are dissipative and there is, therefore, a continuous flux of energy through the systems which maintains their circulations. Because of the dissipation of energy, the memories of the ocean and atmosphere are limited. In the atmosphere the KE can be dissipated in about five days if not replenished from the APE reservoir, whilst in the ocean the KE can be dissipated within 100 days. These limited dynamical memories imply that the predictability of the behaviour of, say, an extra-tropical cyclone or an ocean eddy is limited by the dissipation time scale. It is, therefore, quite impossible to describe the behaviour of the atmosphere or ocean in terms of the influence of an individual eddy circulation. The only possible approach is to evaluate the statistical influence of these highly variable eddies on the behaviour of the general circulation.

To describe the behaviour of the general circulation of the atmosphere, it is necessary to concentrate attention on two quantities:

(i) The energy of the atmosphere and, in particular, the transfer of energy between the equator and the pole.
(ii) The momentum balance of the atmosphere. This is important because it gives an insight into the reasons for the observed distribution of wind, as shown in Figure 6.10, and also into why the sub-tropical jet streams are located at 30° latitude. It also yields information about the role of extra-tropical cyclones in maintaining the climatological wind distribution.

A poleward flux of energy is required by the ocean–atmosphere system to counteract the net radiation imbalance between the poles and the equator. The atmospheric energy, E, can be written as:

$$E = \text{KE} + \text{IE} + \text{PE} + \text{LE} \qquad (9.10)$$

where KE is the kinetic energy, IE is the internal energy, PE is the potential energy and LE is the latent heat of the atmospheric water vapour. The KE of the atmosphere is small compared with the other energy terms and it is generally ignored in the total energy budget. The IE and PE are usually grouped together as the potential heat flux.

The change in the energy of the atmosphere with time, dE/dt, is given by:

$$\frac{\mathrm{d}E}{\mathrm{d}t} = Q - D \qquad (9.11)$$

where Q is the energy source and D is the energy dissipation by frictional processes. The diabatic heating term, Q, is composed of the net radiation heating, the sensible heat flux from the surface and the latent heat released by condensation. For the atmosphere as a whole, the dissipation, D, must balance the net heat input, Q. However, between the equator and the pole, the equation does not balance at every point, and therefore a transfer of energy is required to maintain the distribution of heat sources and sinks. Figure 9.11 shows the meridional distribution of Q in the Northern Hemisphere winter. Everywhere in the troposphere the atmosphere is cooled by long-wave radiation at a rate of between 1 and 2 K day^{-1}. Sensible heat flux from the surface heats the lower 1–3 km of the troposphere, especially in the winter hemisphere, and latent-heat release is responsible for heating all of the troposphere in the equatorial zone, and for heating the lower and middle troposphere at higher latitudes. The net heating distribution shows the whole equatorial zone and the surface boundary layer as net heat sources, whilst the remaining regions are net heat sinks. To maintain this distribution, energy has to be transferred both polewards and upwards by the general circulation. The horizontal flux of energy, F_A, across a latitude circle is given by:

$$F_A = \frac{2\pi a \cos\phi}{g} \int_{p_0}^{p_s} [\bar{v}\bar{E}] \, \mathrm{d}p \qquad (9.12)$$

where v is the northward velocity, E is the energy and $2\pi a \cos\phi$ is the distance along a latitude circle. The bar denotes a time average and the square brackets denote a zonal average.

Figure 9.11 Components of diabatic heating for the atmosphere for December–February. Units: degrees per day

The mean northward transport of energy can be partitioned into two terms:

(i) $\bar{v}\bar{E}$ – the transport of mean energy by the average meridional circulation such as the Hadley and Ferrel cells.
(ii) $\overline{v'E'}$ – the transport by the fluctuating component of the meridional circulation, or the eddy flux. This term includes fluctuations caused by the day-to-day variations in the position

of the Rossby waves in the upper troposphere and by synoptic-scale motions in the lower troposphere.

Figure 9.12 shows the contributions of the potential heat flux and the latent heat flux to the total heat transport. It can be seen that the mean meridional circulation is very effective at transferring heat at low latitudes, but it is less efficient in middle and high latitudes. The Ferrel cell in the middle latitudes actually transfers heat towards the equator against the temperature gradient. At middle and higher latitudes, all of the poleward heat flux is achieved by the eddy circulations. These eddy circulations have not only to transfer the heat demanded by the distribution of heat sources and sinks, but they also have to transfer an extra amount to counterbalance the equatorward transport of heat by the mean circulation. The latent heat flux shows a different behaviour from that of the potential heat flux. At low latitudes, the Hadley cell transports water vapour towards the equator where it is required because of the net deficit of water in the rising

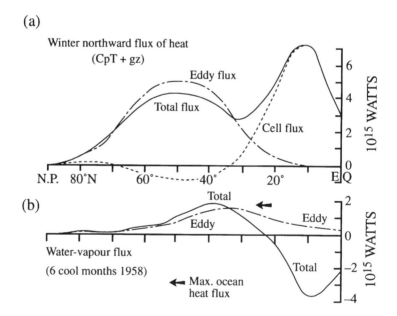

Figure 9.12   (a) Northward heat flux, and (b) water-vapour flux in Northern Hemisphere winter. Arrow shows latitude and magnitude of strongest northward heat flux by ocean currents (an annual average)

branch of the Hadley cell, as described in Section 5.4. At higher latitudes, the eddy flux of water vapour, like the eddy heat flux, is much larger than the mean meridional circulation. However, because the majority of the water vapour is confined to the lowest 5 km, the Ferrel circulation actually produces a latent heat flux in the same direction as the eddy transport. In addition, the major contribution of the eddy latent heat flux is made by the synoptic-scale motions in the lower atmosphere. The total transport of energy by the atmosphere necessary to balance the distribution of heat sources and sinks can be obtained by the summation of the potential heat flux and the latent heat flux.

In conclusion, it has been shown that the variability of the atmosphere is not purely random but that it acts in a systematic manner to transfer energy from low to high latitudes, and eddy circulations play an important role in this process. A similar approach is also necessary to explain the momentum balance of the atmosphere.

The total angular momentum of the Earth, including the atmosphere and the ocean, is constant. Any changes in the angular momentum of one component of the system must be balanced by a corresponding change in the angular momentum of another of the system's components. The atmosphere exchanges angular momentum with the solid Earth by the frictional torque that it exerts on the Earth's surface. If this net torque were in the direction of rotation, then the Earth would increase its angular momentum, and hence its rotation rate, at the expense of the atmosphere's angular momentum. Correspondingly, a net torque in the opposite direction would slow the Earth's rotation rate and the atmosphere would gain angular momentum. Although small variations in the angular velocity of the Earth do occur, it can be assumed that, over a period of one to two years, the angular velocity is constant. Hence, the total angular momentum of the atmosphere should be in balance when it is averaged over several seasonal cycles.

The net torque on the Earth's surface does vary with latitude. At low latitudes the winds have a slower rotation rate than the Earth, and therefore the surface torque acts in the opposite direction to the Earth's rotation. The Earth will lose angular momentum and the atmosphere will gain eastward angular momentum at these low latitudes. In the middle latitudes the surface winds have a higher rotation rate than the

Earth and they produce a torque in the same direction as the Earth's rotation. In this case, the atmosphere will lose and the Earth will gain angular momentum. There will, therefore, be a sink of atmospheric angular momentum in the middle latitudes and a source in low latitudes. In order to maintain this distribution of sources and sinks, the meridional component of the atmospheric circulation must transport angular momentum from low to middle latitudes. Over the polar regions the surface winds are westwards and these areas therefore act as a small source of angular momentum which will also be transferred to middle latitudes.

The angular momentum per unit mass of a parcel of air is given by:

$$a \cos \phi u + a^2 \cos^2 \phi \Omega \qquad (9.13)$$

where u is the zonal component of the wind, Ω is the rotation rate of the Earth, a is the radius of the Earth and ϕ is the latitude.

The first term is the relative angular momentum of the zonal wind and the second term is the angular momentum of the Earth. Furthermore, the annual average of the northward flux of angular momentum, $M\phi$, integrated over the depth of the atmosphere, can be shown to be:

$$M\phi = \frac{2\pi a \cos \phi}{g} \int_0^{p_0} [\overline{uv}] \, a \cos \phi + \Omega \, [\overline{v}] \, a^2 \cos^2 \phi \, dp \qquad (9.14)$$

The overbar indicates a time average, the square brackets indicate a zonal average and v is the meridional component of the wind.

It will now be shown that the total transport of angular momentum consists only of the first term in equation 9.14. Consider a column of the atmosphere extending around a latitude circle. A northward mass transfer into the column at one level must be balanced by an equal southward mass transfer out of the column at a different level in order to avoid a net accumulation of mass in the column. Hence, the total mass transport over the whole atmospheric column is zero and integration of the zonal average meridional velocity $[\overline{v}]$ over the column is also zero. Therefore:

$$M\phi = \frac{2\pi a^2 \cos^2 \phi}{g} \int_0^{p_0} [\overline{uv}] \, dp \qquad (9.15)$$

Now the term $[\overline{uv}]$ is the time-averaged correlation between the zonal wind and the meridional wind. It can be partitioned into two components in a similar manner to the potential heat and latent heat water vapour flux. The first component is the contribution of the mean circulation, $[\bar{u}][\bar{v}]$, and the second is the contribution of the eddy circulation, $[\overline{u'v'}]$, where u' and v' are the departures from the mean wind. The northward transport of angular momentum from low to middle latitudes, shown in Figure 9.13, which is required to maintain the surface torque balance is dominated by the eddy flux, $[\overline{u'v'}]$, and it is only in equatorial latitudes where the Hadley cell contributes an equivalent mean transport. The Hadley cell produces a positive contribution to the $[\bar{u}][\bar{v}]$ term, since the poleward branch of the cell (i.e. $[\bar{v}] > 0$) is associated with an eastward wind (i.e. $[\bar{u}] > 0$), whilst the low-level equatorward flow (i.e. $[\bar{v}] < 0$) has westward winds for which $[\bar{u}] < 0$. Therefore, the product $[\bar{u}][\bar{v}]$ is positive in both cases and the Hadley cell is responsible for a poleward transport of angular momentum.

The Ferrel cell produces a small equatorward transport of angular momentum and therefore, as with the potential heat flux, the eddy momentum flux is the only mechanism available to produce the required poleward momentum flux. The maximum eddy momentum

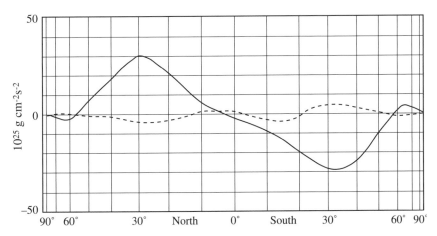

Figure 9.13 The transport of angular momentum by eddies (solid curve) and by the mean meridional circulation (dashed curve). Northward transport is positive

flux occurs in the upper troposphere a few degrees equatorwards of the sub-tropical jet streams. It is this convergence of the eddy momentum flux which provides the source of angular momentum to maintain the jet streams.

Although the meridional cells are not the major contributors to the meridional flux of angular momentum, they are important in the *vertical exchange* of momentum. The Earth's angular momentum, represented by the 'Ω' term in equation 9.14, is a maximum at the equator and decreases towards the poles. The ascending branch of the Hadley cell has a larger angular momentum than the descending branch at a higher latitude. The Hadley cell will therefore produce a net upward flux of eastward momentum. The Ferrel cell acts in the opposite direction and brings eastward momentum down to the surface. Vertical eddy fluxes, associated with large convection cells in low latitudes and synoptic-scale systems in higher latitudes, will also contribute to the vertical flux of angular momentum.

It is seen, therefore, that both eddy fluxes and the mean meridional air circulations contribute to the horizontal and vertical transfer of angular momentum necessary to maintain the observed distribution of surface wind. The eastward winds at middle latitudes and westward trade winds at low latitudes are a necessary consequence of the conservation of angular momentum. Furthermore, because the torque is given by $a \cos \phi \tau_0$, where τ_0 is the surface wind stress and ϕ is the latitude, the eastward winds at middle latitudes must generally be stronger than the westward winds at low latitudes in order to maintain the torque balance.

Recent studies on the momentum balance of intense current systems in the *ocean*, such as that of the Gulf Stream, have also shown that eddy circulations are very effective at pumping angular momentum from the surface layers into the deep ocean. The addition of momentum to these deeper regions produces well-organized, quasi-permanent deep gyre circulations under the Gulf Stream which are larger than the individual mesoscale eddies. Therefore, in both the ocean and the atmosphere, the transient eddies not only contribute to the exchange of heat and other tracers, but are also instrumental in maintaining the observed mean circulation.

9.5   Models of the atmosphere and the ocean

A meteorologist or a physical oceanographer is at a distinct disadvantage compared with scientists in many other disciplines for he or she has only one laboratory and that is the atmosphere or the ocean! Experimental programmes carefully designed to study a particular scale of motion, such as a tornado or a mesoscale eddy, are themselves exposed to the vagaries of the atmosphere and ocean. Often, the measurements obtained are incomplete and definitive conclusions cannot always be drawn. A well-designed current meter array may be rendered useless by the loss of moorings in a strong current or a storm. Despite improvements in the reliability of technology, the lack of control over experimental conditions is a unique factor of the science of meteorology and oceanography.

This basic experimental difficulty has encouraged the use of laboratory models of both the atmosphere and the ocean. These models are of two types. The first type is a physical analogue model. For instance, an analogue of the atmospheric general circulation can be obtained by rotating an annulus containing a fluid on a turntable. The temperature gradient between the pole and the equator is simulated by heating the outer rim of the annulus and by cooling its inner rim. The experiment facilitates the study of the behaviour of the flow under controlled conditions. The experimental scientist is able to change the rotation rate and temperature gradient at will, and is able to repeat the experiment many times to check the validity of the results.

The second type of model is the numerical analogue which uses the computer. These models are based on the numerical solution of a set of differential equations which approximate closely to the equations of motion on a rotating Earth (i.e. equations 7.17 and 7.18). These models can be used to study a variety of scales of motion, from the general circulation of the atmosphere and ocean to the detail of a thunderstorm or tornado. Again, the experimenter has the advantage that the parameters in the model can be changed at will in order to gain a deeper insight into the mechanisms of the phenomena under study. Furthermore, the modeller has a complete set of calculated data from which, for example, the energy and momentum balance of the system can be deduced. Models of large and synoptic scales of the atmosphere are now used routinely for weather prediction for up to 10 days ahead. General circulation models, which include, in addition to the equations of motion, numerical analogues of physical processes such as radiation absorption and emission, clouds and surface processes, can be used to study the longer-term effects of sea surface temperature anomalies, volcanic aerosols and changes in the atmospheric constituents, such as carbon dioxide. In the last two decades the numerical model has indeed become a very powerful tool for understanding the behaviour of the atmosphere and ocean.

A simplified numerical model of the atmosphere will now be considered. It is of the type used for weather prediction and is presented in order to illustrate the basic procedures. The atmosphere is first gridded in three dimensions. There are usually between five and 15 levels in the vertical dimension, and the majority of these levels are in the troposphere. In the horizontal dimension the resolution has to be between 100 and 200 km in order to resolve synoptic-scale circulations such as cyclones and anticyclones. The observed state of the atmosphere at time t is obtained from synoptic weather stations using radiosonde balloon measurements. This information, which includes the temperature, humidity, pressure and the horizontal components of the wind, is then interpolated onto the three-dimensional grid. For a 100 km grid covering the Northern Hemisphere with 10 levels, about 2.5×10^6 numbers are required to describe the initial state of the atmosphere. The x component of the equation of motion can be written as:

$$\frac{\partial u}{\partial t} = F(u, v, w, p) \qquad (9.16)$$

where $\partial u/\partial t$ is the local acceleration of velocity. The right-hand side includes the accelerations arising from the pressure gradient, the Coriolis force and inertial forces. From the initial data, all these accelerations can be obtained by numerical approximation. For example, the pressure-gradient acceleration, $(1/\rho)(\partial p/\partial x)$ at a point i, can be approximated by $(1/\rho_i)[(P_{i+1} - P_{i-1})/2\Delta x]$, where Δx is the grid spacing. The net acceleration, F, can thus be obtained at each grid point at time t. Furthermore, by numerical approximation, equation 9.16 can be written as:

$$\frac{u(t + \Delta t) - u(t)}{\Delta t} = F(t)$$

Hence:

$$u(t + \Delta t) = u(t) + \Delta t F(t) \qquad (9.17)$$

As $F(t)$ and $u(t)$ are known, u can be calculated at a later time, $t + \Delta t$. The time step, Δt, would usually be between 10 and 30 min. A similar procedure can be applied to the y component of motion and to the other equations for the vertical motion, temperature, pressure and humidity to obtain values of all the variables at the later time $t + \Delta t$. By repeated application of equations similar to equation 9.17, values at $t + 2 \Delta t, t + 3 \Delta t$, etc., are thus obtained. For a time step of 10 min, 144 cycles would be necessary to obtain a forecast for one day ahead.

In principle, it would seem that the calculation could be extended indefinitely into the future to give forecasts of not only a few days ahead, but for a month or even a year ahead. However, there are both practical and theoretical limitations to prediction. The first limitation lies in the errors in the prescription of the initial state of the atmosphere. These errors are associated with interpolation of data onto the grid, particularly in regions of sparse data such as over the ocean. The second limitation is that the grid can only resolve scales of motion which are larger than the grid spacing, and therefore systems such as squall lines, thunderstorms and cumulus convection cannot be included explicitly in the prediction. The non-linearity of the atmosphere implies that these unresolved scales may transfer energy into the larger-scale systems and thus change the development of a synoptic system. Lastly, the atmosphere is dissipative, and therefore it cannot retain information of the previous history of the atmospheric state for more than about 10–14 days into the future. Hence, all numerical predictions will degrade as the integration proceeds.

General circulation models of the atmosphere can be integrated for much longer periods than the dissipative time scale, ranging from months to many years. Here, the interest is not in the prediction of an individual cyclone or anticyclone, but in the long-term statistics of the flow. The information may be time averaged to produce maps of surface pressure, velocity and temperature similar to standard climatological maps shown in Figures 6.10 and 6.11. In addition, the momentum balance and energy balance may also be analysed for comparison with observed estimates. This type of model is very much a research tool for understanding the behaviour of the atmosphere, but it is anticipated that, eventually, these models will be used to make monthly and seasonal predictions. These predictions will inevitably be different from short-term weather forecasts because they will be statistical rather than deterministic in nature.

Ocean circulation models are based on a similar set of dynamical equations to those used in atmospheric circulation models. The equations are discretized into a set of boxes of uniform horizontal dimensions, but variable thickness in the vertical (see Figure 9.14). The horizontal flow is predicted by the momentum equation, which includes the horizontal pressure gradient, Coriolis acceleration and forcing terms, due to the surface wind stress and dissipation terms. The vertical motion is obtained from the horizontal divergence of the flow.

Surface buoyancy fluxes due to heat and fresh water will change the temperature and salinity in the surface layer of the ocean. However, the flow will carry these properties far into the interior of the ocean where they will mix with other water masses.

These processes of transport, and mixing of temperature and salinity, are described by a temperature and salinity equation for each ocean box in the model. From the equation of state for sea water, the density can be calculated and, from this, the hydrostatic pressure in each box is obtained. The horizontal pressure gradient can then be determined for the momentum equation. This set of time-dependent equations can then describe all the dynamical components of the flow field. The model is usually started from a state of rest of uniform salinity and temperature, and is spun up by surface wind and buoyancy forcing. In order to reduce the amount of time to produce a realistic ocean circulation, the climatological ocean temperature and salinity may be used to constrain the model. Despite these constraints, these models are able to produce many features of the observed eddy fields. We have seen earlier that the energetic scales of motion for the ocean are of the order of tens to hundreds of kilometres, which are an order of magnitude smaller than their atmospheric counterparts. In order to resolve this motion we have to have horizontal boxes of the order of 10 km square compared with 100 km square in the global atmospheric models. This means that a hundred more calculations are required to cover a similar area to an atmospheric model. Furthermore, the time

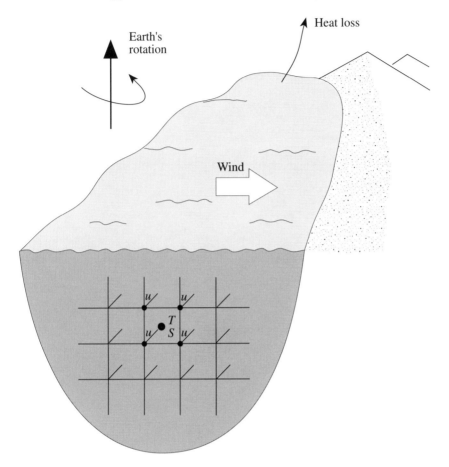

Figure 9.14 A schematic diagram of an ocean general circulation model. The ocean is partitioned into a regular lattice and equations for velocity (u) are solved at the corners, whilst the temperature (T) and salinity (S) equations are solved at the centres of the lattice. The model is either (a) forced by climatological wind stress, surface heat and freshwater fluxes, or (b) coupled to an atmospheric model which predicts these fluxes. Reproduced with permission of Mason Publishing

step, Δt, has to be reduced as the resolution is increased to maintain numerical stability of the calculation. For these reasons, the global ocean circulation models require about a thousandfold more computer power than their atmospheric equivalents. Global ocean models of this resolution are now being evaluated using data collected from the World Ocean Circulation Experiment (WOCE).

For present climate prediction studies, much lower resolution ocean models which parameterize the eddies as a sub-grid-scale process are used. This is because the eddy models are very computationally expensive and currently can only be run for periods of 10 to 20 model years.

10

Climate Variability and Predictability

10.1 Air–sea interaction: an introduction

Air-mass transformation over the ocean

The modification of a continental air mass as it moves over the ocean is one of the more important processes necessary to the understanding of the interaction between the ocean and the atmosphere. In order to illustrate these processes, two examples of air-mass transformation will be considered.

The first example is the transformation of the trade winds on their journey from the sub-tropical land masses to the ITCZ. Figure 10.1 shows a schematic diagram of some of the salient features of the trade-wind system between the north-west African coast and the equatorial Atlantic Ocean. Dry, warm surface air from the continent is rapidly transformed by contact with the cooler surface water in a shallow boundary layer. The moisture content in the boundary layer is rapidly increased by evaporation from the ocean, whilst sensible heat is lost from the air to the ocean because of the reversal of the vertical temperature gradient. In addition, the air is cooled by the emission of long-wave radiation into space. The surface air therefore cools rapidly to the sea surface temperature and, because of the addition of water vapour and the net cooling, low stratus clouds form in the boundary layer. At this stage the boundary layer may be relatively shallow, having a depth of a few hundreds of metres. Above the maritime boundary layer the air is potentially warmer and very dry because of the descent of air from high levels in the troposphere associated with

Figure 10.1 Schematic diagram of the modification of the trade-wind boundary layer over the ocean

subsidence in the tropical anticyclones. Therefore, at the boundary between the two air masses, a strong trade-wind inversion is formed which tends to suppress vertical motion and mixing. As the maritime air moves over progressively higher sea surface temperatures, the latent heat flux increases and the sensible heat flux becomes upwards from the ocean to the atmosphere. The warming of the air mass results in a break-up of the stratus cloud and the development of trade-wind cumulus clouds. The shallow cumulus convection mixes the moist surface air with the dry air above the inversion. This results in a deeper boundary layer. The convection also brings drier air towards the surface which, in turn, increases the evaporation from the ocean into the boundary layer. In the western tropical ocean basins the boundary layer reaches a depth of 2–3 km and evaporation is three times larger than in the eastern ocean basins. In these boundary layers, convection is deep enough to produce showers and this warms the cloud layer by latent heat release. Energy budgets of the trade-wind boundary layer have shown that the downward heat flux through the inversion is as large as the upward surface flux of sensible heat. They have also shown that virtually all of the water evaporated into the boundary layer is exported horizontally out of the trade-wind zone and is subsequently released in the tropical convergence zones. A typical trade-wind velocity is $6 \, \mathrm{m \, s^{-1}}$, and therefore the surface air mass, providing that it remains in the boundary layer, will traverse a distance of 2500 km between the source region and the equatorial region in about five days.

The second example of air-mass transformation is more typical of air–sea interaction in middle latitudes. During winter very cold, dry air from the northern Asian and North American land masses periodically surges out over the relatively warm waters of the western North Pacific and North Atlantic Oceans. These cold outbreaks occur in the north-westerly flow in the rear of an extra-tropical cyclone and they may last for several days. Figure 10.2 shows the transformation of a cold, dry stable air mass as it flows out over the East China Sea and the warm Kuroshio current. Initially, the continental air mass may have a temperature between 0 and -20°C, and a very low specific humidity of less than $1\,\mathrm{g\,kg}^{-1}$. The sea surface temperature, in contrast, increases from about 5°C near the coast to 20°C over the Kuroshio current. As the air mass first moves out across the East China Sea, the unstable boundary layer will be rapidly warmed by an upward flux of sensible heat. Although the air is dry, it is initially very cold and is therefore unable to absorb much moisture. As the boundary layer warms and deepens, the surface latent heat flux becomes comparable with the sensible heat flux. In contrast to the trade-wind boundary layer, which deepens only slowly, these cold air outbreaks cause rapid changes in the depth of the boundary layer. Over a spatial scale of $300\,\mathrm{km}$ depth, changes of 1–$2\,\mathrm{km}$ are observed. During the Air-Mass Transformation Experiment (AMTEX) in 1974 and 1975, total surface heat fluxes of 700–$800\,\mathrm{W\,m}^{-2}$ were observed over two separate four-day periods. This heat flux, which is half the solar constant (i.e. $1360\,\mathrm{W\,m}^{-2}$), would heat a column of air $2\,\mathrm{km}$ deep at a rate of $30\,\mathrm{K\,day}^{-1}$ and would lead to vertical instability of the boundary layer. The growth of convective cloud is limited by the height of the inversion which marks the boundary between the modified air mass below and the continental air mass above. Near the coast the boundary layer depth is small, and therefore the convective cloud is shallow, being usually stratocumulus and of small horizontal dimensions. However, as the boundary layer grows, both the depth and horizontal scale of the convective cloud increases, and it can usually grow sufficiently to produce showers. The regions of large air-mass transformation are also regions of frequent cyclonic development, as the result of the larger gradient of surface temperature. The region south of Japan is the most active region of cyclone generation in the world during the winter season. The average heat loss of the ocean over the Kuroshio current in February is $350\,\mathrm{W\,m}^{-2}$, which is sufficient to cool the upper $100\,\mathrm{m}$ layer of the ocean by 2.2°C in one month.

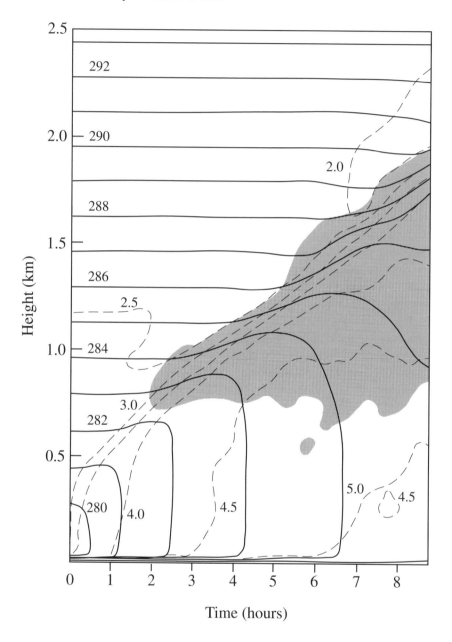

Figure 10.2   Simulation of the evolution of a moist convectively mixed layer in a cold outbreak over the Kuroshio current. Solid lines indicate potential temperature, K, dashed lines specific humidity, $g\,kg^{-1}$, and stippled area is cloud layer

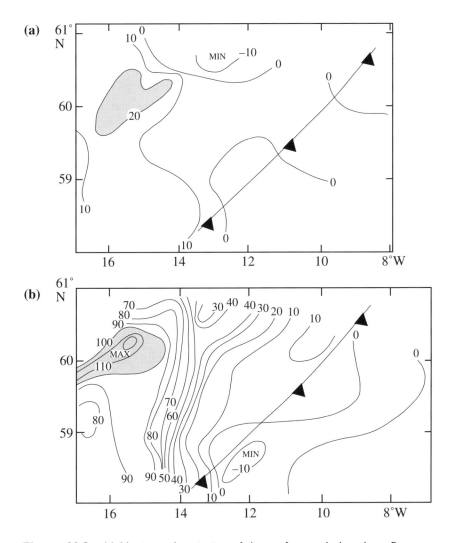

Figure 10.3 (a) Horizontal variation of the surface turbulent heat flux over the ocean during the passage of a cold front (Q_H: W m^{-2}) at 01.00 hours GMT on 31 August 1978. Values of more than 20 W m^{-2} are shown hatched. (b) Horizontal variation of the surface latent heat flux (Q_E: W m^{-2}) at 01.00 hours GMT on 31 August 1978. Values of more than 100 W m^{-2} are shown hatched

The two previous examples have shown that the extraction of heat from the ocean is strongly dependent on the temperature and humidity characteristics of the overlying air mass relative to the sea surface temperature. In an extra-tropical cyclone, two or three different air masses may cross a given area in a day and produce large variations in surface heat fluxes. Figure 10.3 shows the surface fluxes of latent and sensible heat during a period of relatively weak air–sea interaction in summer in the north-eastern Atlantic Ocean. The cold front marks the boundary between a cooler, drier air mass to the north-west and a warm air mass to the south-east. In the warm sector, both latent heat and sensible heat fluxes are close to zero, and the air mass is therefore receiving little or no heat from the ocean. This is typical for warm, humid air masses moving over progressively lower sea surface temperatures. In the rear of the cold front, the latent heat flux increases to a maximum of $110 \, \text{W} \, \text{m}^{-2}$ about 300 km from the front, whilst the sensible heat flux is much less, being in the range from 10 to $20 \, \text{W} \, \text{m}^{-2}$. This contrasts strongly with the cold outbreaks over the Kuroshio current where sensible heat fluxes are of a similar magnitude to latent heat fluxes. It will also be noticed that, in the rear of the cold front, there are variations in latent heat flux caused by horizontal variations in the sea surface temperature of 1°C.

Response of atmospheric phenomena to sea surface temperature

All scales of atmospheric motion are involved in the transfer of energy between the ocean and atmosphere, and therefore most atmospheric phenomena will respond, directly or indirectly, to the sea surface temperature. On the small scale, an increase in sea surface temperature may change the stability of the overlying air mass which may, in turn, change the boundary layer cloud from a stable stratus type to an unstable convective type. Similarly, the distribution of sea fog may be affected by a relatively small change in the sea surface temperature. Mesoscale circulations may also be influenced by the sea surface temperature. The intensity of a sea-breeze circulation depends on the difference in temperature between land and sea to generate the available potential energy which sustains the circulation. Mesoscale convection over the ocean will depend on the air–sea temperature difference to produce latent and sensible heat fluxes to sustain the vertical circulation. In turn, the vertical circulation will bring colder and drier air to the boundary layer, and this will enhance the vertical transfer of heat into the circulation.

The tropical cyclone is an example of the self-sustaining circulation described above. The generation areas of the tropical cyclone are usually limited to the regions of the tropics where sea surface temperatures are above 26ºC. Evaporation rates are determined principally by the surface wind and by the sea surface temperature. It is also noted that the vapour pressure has a non-linear dependence (Figure 4.9), and thus high evaporation rates occur over the tropical ocean. An embryonic low-pressure system will produce a convergence of water vapour in the maritime boundary layer which will release latent heat when uplifted. As the horizontal and vertical circulation gains kinetic energy from the latent-heat release, the surface evaporation rate and the surface convergence of water vapour will both increase, and the circulation will intensify. Provided that there is a copious supply of water vapour, the storm will increase in intensity at an exponential rate until the loss of kinetic energy by surface friction balances the kinetic energy input. A mature tropical cyclone can produce surface latent heat fluxes of $1000 \, W \, m^{-2}$ and sensible heat fluxes of $500 \, W \, m^{-2}$. These are approximately twice the surface heat flux observed in cold outbreaks over the Kuroshio current. The tropical cyclone is a truly oceanic phenomenon because, when it strikes land, its surface moisture source is severely reduced and this, in combination with higher surface friction, will cause it to dissipate quickly. Occasionally, tropical cyclones will propagate into higher latitudes where they may trigger the development of an extra-tropical cyclone. Over the higher-latitude oceans, the extra-tropical cyclone may increase its energy by ventilating the marine boundary layer with drier air which has subsided from the upper troposphere in the rear of the cold front. The dry air is able to pick up heat and moisture which can both be recirculated into the ascending region of the cyclone. They may be released subsequently as available potential energy and kinetic energy.

The larger-scale circulation of the atmosphere is strongly influenced by the distributions of the continents and oceans. In middle latitudes the climatological position of the long Rossby waves is determined, in part, by the distribution of heat sources and sinks. Figure 10.4 shows the observed surface pressure and 50 kPa height, as well as surface heat sources and sinks at middle latitudes of the Northern Hemisphere. The principal feature predicted by theory (Smagorinsky, 1953) is the development of a surface anticyclone (surface cyclone) about 25º

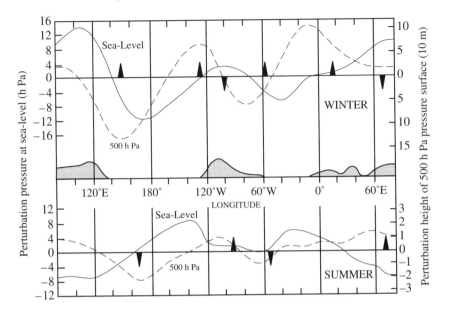

Figure 10.4   Zonal profiles of normal perturbation pressure at sea level (solid line) and normal perturbation height of the 50 kPa (500 hPa) surface (dashed line). Continental elevations are indicated by a cross-hatched area. Positions of observed relative heat sources and sinks are indicated by spires pointed upwards and downwards, respectively. Upper part: winter profiles averaged over 20° latitude centred at 45°N. Lower part: summer profiles averaged over 20° latitude centred at 50°N

downwind of the heat sink (heat source) and a ridge (a trough) at 50 kPa displaced by 10° to the west. During winter, the North Atlantic and North Pacific Oceans provide an average surface heat source of 100 W m⁻². The largest heating is located over the Kuroshio and Gulf Stream currents, as seen in the previous sub-section. In contrast, the continental land masses are net heat sinks in winter because heat loss by long-wave radiation exceeds the input by solar radiation. According to theory, surface low-pressure centres should be expected about 25° downstream of the heat sources in the north-western Atlantic and north-western Pacific. This is in rough agreement with the observed positions of the Icelandic and Aleutian lows. Over the Eurasian continent, the Siberian anticyclone should be displaced towards the east of the continent. This is once more in rough agreement with Figure 10.4. However, the anticyclone over North America is found displaced towards the west rather than the east. The reason for this is the effect of the Rocky Mountains which not only alters the relative position of the heat sources and sinks, but also compresses the atmospheric column, as discussed in Section 7.4, to produce anticyclonic vorticity over the mountains themselves. Downstream of the Rocky Mountains the stretching of vortex tubes causes the formation of a trough in the upper air flow. Recent computer models have shown that the climatological atmospheric flow is determined by both large mountain barriers and by the distribution of heat sources and sinks. The upper tropospheric flow is most sensitive to the distribution of topography, whilst the flow in the lower troposphere is determined by the pattern of continental and oceanic heating. However, the presence of mountain barriers, such as the Himalayas, impedes the transport of moisture in the lower troposphere, and thus indirectly influences the distribution of heat sources.

In tropical latitudes the differential heating between the Asian land mass and the relatively cool Indian Ocean is responsible for the energy of the south-west monsoon circulation. These monsoon circulations are associated with a large-scale, interhemispheric transfer of latent heat from the southern Indian Ocean and Arabian Sea which, when released over the Indian sub-continent, strengthens the monsoon circulation. The vertical flux of latent and sensible heat from the Arabian Sea during the onset of the south-west monsoon in June has a typical value of $200\,\mathrm{W\,m^{-2}}$. This large heat loss produces a rapid cooling of the ocean mixed layer of $2\,^{\circ}\mathrm{C}$ in one month and a deepening of the oceanic mixed layer from $40\,\mathrm{m}$ in early June to $100\,\mathrm{m}$ in early July.

The above examples illustrate some of the complex interconnections which exist between the ocean and atmosphere. The ocean is able to store vast quantities of heat during the inactive season and then release it into the atmosphere, when required, to enhance and to ultimately drive the atmosphere on a variety of scales, from the mesoscale and synoptic-scale circulations to the large-scale monsoon circulations.

10.2   Seasonal anomalies of the ocean–atmosphere system

The ocean plays a major role in the seasonal changes of the atmospheric circulation, both in the tropics and at higher latitudes. It is therefore reasonable to assume that anomalies from the seasonal cycle may be linked to anomalous air–sea interactions. Before considering the possible mechanisms of these interactions, an example of anomalous behaviour of the atmosphere and ocean in the North Atlantic Ocean between January and June 1972 will be considered. Figure 10.5 shows both the atmosphere and sea surface temperature anomalies for June 1972. Both patterns are representative of the chosen period between late winter and early summer. The anomaly patterns are obtained from the differences between the observed pattern and the long-term climatic normals for June. The most notable features are:

(i)   The negative temperature anomalies, up to 3ºC in the north-eastern North Atlantic, and the positive temperature anomalies, up to 5ºC in the western and central Atlantic at 40ºN, as well as smaller negative temperature anomalies at lower latitudes.

(ii) An anomalous cyclonic circulation over the British Isles with a cold north-westerly flow over the northern North Atlantic, and an anomalous anticyclonic circulation over the central Atlantic Ocean. The cold north-westerly winds of this year produced a cold wet spring and early summer in north-western Europe, and were associated with the largest volume of ice recorded in the Labrador current since 1912, when the *Titanic* sank. In June 1972 an extremely active hurricane, called Agnes, followed an unusual northerly path along the east coast of the USA into the Great Lakes region. It is thought that the anomalous southerly wind off the east coast of the USA and the high sea surface temperatures prevailing there were responsible for the behaviour of Hurricane Agnes.

From Figure 10.5 it can also be seen that the patterns of sea surface temperature anomalies and atmospheric circulation have similar horizontal scales, and that they are clearly related. The cold northerly atmospheric flow is associated with negative sea surface temperature anomalies, whilst the warm southerly flow to the west of the anticyclone occurs over the positive sea surface temperature anomaly. Thus, the atmospheric flow is tending to maintain the sea surface

Figure 10.5   (a) Sea surface temperature anomalies (°C) for 1–26 June 1972. (b) Atmospheric circulation anomalies at 70 kPa height for June 1972. Intervals are in units of one standard deviation; maxima and minima are labelled. Zero isopleth is the heavy line. The arrows indicate the direction of the atmospheric circulation

(a)

(b)

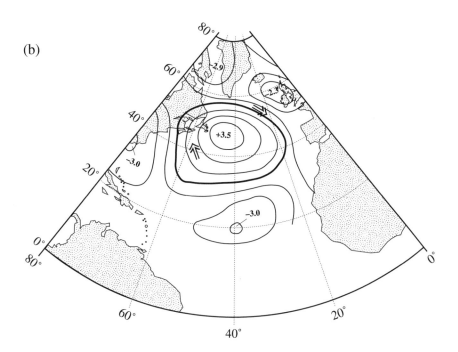

temperature anomaly patterns by extracting more heat from the northern North Atlantic Ocean than is normal and by extracting less heat than usual from the western Atlantic. In turn, the large heat flux from the ocean in the northern North Atlantic Ocean would provide an anomalous source of heating which would tend to maintain a low-pressure system downstream of the heat source. Conversely, the reduced heat flux in the western Atlantic would produce an anomalous heat sink which would maintain an anticyclone in the central Atlantic. These arguments are not necessarily conclusive because large-scale interactions with the tropical atmosphere and the Pacific Ocean may also be responsible for the anomaly pattern, but they do suggest that large-scale air–sea interaction can produce a persistent, anomalous pattern which may be reinforced by heat exchange between the ocean and the atmosphere.

In the above example the ocean is behaving as a slave to the atmosphere but, because of its heat storage capacity, it is able to maintain anomalous patterns of heating and cooling. Without the ocean the atmosphere would lose all of its memory of the thermal anomalies within a period of a few weeks. The longevity of the above type of air–sea interaction has aroused much interest in the possibility of using these patterns for long-range seasonal forecasting. However, progress in this area has been limited. Recent studies have shown that the predictability of these air–sea interaction events is limited by the predictability of the atmosphere – which is generally less than two weeks. However, once an air–sea interaction event has developed, the ocean will tend to maintain the pattern for a period of one to a few months. Thus, the atmosphere initiates the anomaly, whilst the ocean tends to make it persist. Ultimately, the seasonal cycle limits the longevity of these air–sea events by changing the distribution of heat sources and sinks. This eventually results in a breakdown of the anomaly pattern.

An interesting example of the effect of the seasonal cycle on air–sea interaction events has been described by Namias (1971). In the north-eastern Pacific Ocean, it has been found that warm sea surface temperature anomalies, which develop in the summer months, tend to persist through the autumn and into winter. During the autumn, the extra-tropical cyclones intensify in response to the increase in the poleward temperature gradient and the differential heating between ocean and continent. The presence of a warm sea surface temperature anomaly increases the heat available, via latent heat and sensible heat fluxes. Thus, the energies of the storms are also increased and the cyclones tend to become more intense than is normal during the autumn and winter months. This type of air–sea interaction is, in the end, self-destructive because, once the anomalous supply of heat in the ocean mixed layer has been exhausted, the anomalous atmospheric circulation can no longer be maintained. However, it does illustrate the point that the amount of heat stored in the ocean mixed layer during the summer heating season is an important factor for the prediction of air–sea interaction events in the subsequent autumn and winter seasons. A summertime sea surface temperature anomaly of 1ºC, extending to a depth of 50 m, contains approximately $2 \times 10^8 \, \mathrm{J \, m^{-2}}$ of excess energy. If this energy is released over the subsequent three months, it will provide an anomalous heat source of about $25 \, \mathrm{W \, m^{-2}}$, compared with the normal autumnal heat loss of $80 \, \mathrm{W \, m^{-2}}$, in the northeastern Pacific.

There is also evidence that positive sea surface temperature anomalies in the Arabian Sea, which develop in the heating season prior to the onset of the summer south-west monsoon, may provide an additional source of latent heat to enhance the monsoonal circulation described in Section 10.1. In the tropical ocean, relatively small sea surface temperature anomalies of approximately 1ºC can produce large anomalies in evaporation because of the non-linearity of the water-vapour pressure with temperature.

In all of the above examples, the interaction of the atmosphere with the ocean has been confined to the mixed layer which occupies the upper 50–200 m of the ocean. As seen in Section 5.2, most of the seasonal storage of heat is confined to this layer, and therefore many short-term or seasonal anomalies can be explained by variations in the depth and temperature of the mixed layer. However, other interactions may also be significant. The ocean transport of heat shows a seasonal variation both in tropical latitudes and middle latitudes, as illustrated by Figure 5.3. In middle latitudes the seasonal transport variation is small in comparison with the storage term, but in the tropics it is larger than the storage term. These seasonal changes in the ocean heat transport at low latitudes are related to the seasonal variations in the trade winds, with strong trade winds enhancing the poleward transport of heat. Model studies have shown that this seasonal response of the ocean is confined to the direct wind-stress-driven flow in the mixed layer, via the Ekman balance. Hence, strong westward trade winds will produce a poleward transport of warm surface water and cooler sub-surface water returns towards the equator for mass balance.

10.3   Interannual fluctuations in the ocean–atmosphere system

The air–sea interaction events discussed in the previous section occur on monthly or seasonal time scales. They are generally connected with the short-term variability of the atmosphere or the seasonal forcing by solar radiation of the atmosphere–ocean system. On the interannual time scale there is no large external forcing (see Chapter 1), and therefore climate variations arise from internal interactions such as those within the ocean–atmosphere system.

The largest signal on interannual time scales is that of the El Niño–Southern Oscillation (ENSO). The El Niño is the ocean component, whilst the Southern Oscillation is the atmospheric component of the interaction. The ENSO events are irregular events occurring between two and seven years, with a mean frequency of about four years. The interaction involves the surface and thermocline waters of the tropical Pacific Ocean with the overlying circulation in the global troposphere. The El Niño is a warming of the equatorial Pacific Ocean, of 2–8°C, which persists for a number of seasons.

The classical El Niño is the warming of the waters adjacent to the coasts of Ecuador and Peru, which is accompanied by torrential rain and a major reduction in the productivity of coastal waters. The warming usually occurs in December or January, and hence the name El Niño or Christchild. In normal years, the coast has cool surface waters due to strong upwelling of thermocline waters to the surface. The thermocline waters originate from the cool northward Humboldt current. This upwelling water, which is rich in nutrients, absorbs large quantities of solar radiation in the surface layers. These conditions are ideal for the growth of phytoplankton, which in turn sustains the productive anchovy fishery. When El Niño occurs, the cool Humboldt waters are replaced by warm equatorial waters which are low in nutrients. This leads to a very drastic reduction in the productivity of these waters. Consequently, El Niño has a major influence on both the fishing and agricultural economies of Peru and Ecuador.

More comprehensive measurements of sea surface temperature have shown the warming spreads eastwards from the international dateline to the South American coast (see Figure 10.6) and up to 20°S. This is an area of the ocean which is equivalent to 10% of the Earth's surface. Hence, a rise in the global surface temperature of a few tenths of a degree can occur during an El Niño event. For comparison, this is similar to the longer-term rise in the global surface temperature during the last 100 years, and therefore has to be taken into account in the analysis of global warming.

We will now consider the influence this warming will have on the overlying atmosphere. The additional thermal energy which is stored in the Pacific Ocean during a strong El Niño event is about 3×10^{22} joules, which is equivalent to about eight times the total available potential energy and kinetic energy of the atmosphere. Furthermore, if all this extra heat was converted into available potential energy at the observed rate of $2.5 \, \mathrm{W \, m^{-2}}$, then it would be sufficient to maintain the atmospheric circulation for about nine months – a time scale which is slightly shorter than the El Niño event. Therefore, this large-scale warming of the tropical Pacific Ocean will have a profound influence on the atmosphere.

The Southern Oscillation is the major response of the troposphere to the anomalous warming in the Pacific Ocean. It was first described by Sir Gilbert Walker, the Director General of the Indian Meteorological Office, earlier this century. He described a large-scale surface pressure pattern which had its dominant signal over the Indian–Pacific Ocean (Figure 10.7). This pressure signal can be defined in terms of the seasonal pressure difference between Indonesia and Tahiti in the south-east Pacific.

The climatological surface pressure gradient between Indonesia and Tahiti forces a steady south-east trade-wind flow from Tahiti towards Indonesia. During El Niño events, the surface pressure over Indonesia is higher than normal, whilst pressure is lower over the south-east Pacific Ocean, which causes a reduction in the pressure gradient between the two regions and a weakening of the south-east trade winds. The Southern Oscillation index is anomalously low. With a cooler than normal east Pacific, the pressure is lower than normal over Indonesia and higher than normal in the south-east Pacific, which drives stronger than normal south-east trade winds. The Southern Oscillation index is high.

Figure 10.6 Sea surface temperature anomalies (°C) in the tropical Pacific Ocean during an El Niño event (December 1982)

Figure 10.7 The correlation of monthly mean sea-level pressure with that of Jakarta, Indonesia. The correlation is large and negative in the South Pacific, and large and positive over India, Indonesia and Australia. This pattern defines the Southern Oscillation. An index of the Southern Oscillation can be defined as the difference in pressure between Darwin (north Australia) and Tahiti in the South Pacific

This pressure signal is associated with major changes in the tropospheric wind circulation and, in particular, the position of the tropical convergence zones. These convergence zones are regions of heavy rainfall due to the uplift of warm moisture-laden air brought in by the ocean trade winds in the lower troposphere. The cooling due to the uplift of air releases large amounts of latent heat of condensation which gives additional buoyancy to the air and further uplift.

Figure 10.8 shows the principal features of the tropospheric circulation over the tropical Pacific Ocean for (i) normal or cold conditions in the east Pacific, and (ii) during an El Niño event.

In normal conditions the main tropical convergence zones are the ITCZ, which extend from Indonesia in the west to Central America in the east between 5 and 10°N, and the South Pacific Convergence Zone (SPCZ), which extends from northern Australia south-eastwards into the central South Pacific. In order to balance the mass of air brought into the convergence zones by the trade winds, there has to be a compensating outflow of air in the upper troposphere. The outflow has a low water content due to its low temperature and pressure in the upper troposphere, and the removal of water from the atmosphere by rainfall. This outflow of air is directed towards the south-east Pacific to complete the circulation cell. The north–south component of the circulation is known as the Hadley Circulation, whilst the east–west component is known as the Walker Circulation. The upper tropospheric air which converges towards the south-east Pacific descends to the lower troposphere. Because of adiabatic warming as the air descends, this region is associated with clear skies and low rainfall. This region, which is sandwiched in between the convergence zones, is known as the 'dry zone'.

In the ocean the trade winds drive cooler water equatorwards in the Humboldt current, and upwelling along the coasts of South America and the equator. This produces a large cool tongue of water in the eastern tropical Pacific Ocean.

During an El Niño event, the cold tongue is replaced by warmer water from the central Pacific Ocean and from north of the equator. In an extreme El Niño event, such as occurred in 1982–83, the warming is sufficient to completely remove the cold tongue. The weakening of the trade-wind circulation is associated with major displacements of the tropical convergence zones from their normal position. In particular, the ITCZ tends to move south to the equator, whilst the SPCZ will

(a)

(b)

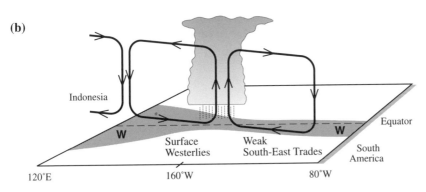

120°E   160°W   80°W

Figure 10.8   The tropical atmospheric circulation over the Pacific Ocean during (a) normal conditions, and (b) El Niño conditions. During normal conditions the surface pressure is low over Australia and Indonesia (high rainfall), and high over the south-east Pacific, so the surface trade-wind circulation is strong and the Southern Oscillation index ($P_{DARWIN} - P_{TAHITI}$, where P is the surface pressure) is high. During El Niño conditions, the pressure is higher over Australia and Indonesia (low rainfall), and lower in the south-east Pacific; consequently, the trade-wind circulation is weaker and the Southern Oscillation index is low (W, warm; C, cold). Reproduced by permission of Manson Publishing

move further east. Over Indonesia and northern Australia, the rainfall is lower than normal and drought conditions may prevail. Conversely, the islands in the central and eastern Pacific, which are normally in the dry zone, will now experience heavy rainfall associated with the displacement of the convergence zones.

Though the major ENSO interaction is predominant in the tropical Pacific Ocean basin, there are influences which extend beyond this region.

The change in the surface tropical pressure associated with the Southern Oscillation has a global influence on the general atmospheric circulation. Its effects cause changes in the extra-tropical circulation, as well as changes in the monsoonal circulations over the Asian and African continents. Indeed, Sir Gilbert Walker's interest in surface pressure patterns was for determining the reasons for the variations in monsoon rainfall from one year to the next over the Indian sub-continent.

During El Niño events, the winter temperatures in western North America are warmer than normal. This is the result of a change in the atmospheric circulation in the middle latitudes of the North Pacific, which in turn is driven by the anomalous warming in the tropical atmosphere over the Pacific Ocean.

Mechanism of the ENSO

It is known that the principal physical processes associated with ENSO are related to the tropospheric wind circulation, the sea surface temperature and the heat stored above the tropical thermocline in the Pacific basin.

During El Niño there is a major redistribution in the heat content in the Pacific Ocean with a transfer of heat towards the eastern Pacific. This heat transfer is the result of the changing patterns of equatorial currents brought about, for example, by the weakening of the south-east trade winds in the eastern Pacific Ocean. Local processes also have a role to play; for example, a reduction in winds will reduce the local upwelling of cooler thermocline waters, which in turn leads to an increase in sea surface temperature. Hence, the wind circulation in the lower troposphere has a crucial role in determining the ocean circulation and redistribution of heat in the ocean.

The sea surface temperature and heat content have a major role to play in the positioning of the major tropical convergence zones, as we have seen in the tropical Pacific Ocean. These convergence zones have a tendency to be located in regions of maximum sea surface temperature. Thus, changes in the location of maximum sea surface temperature will move the convergence zones and alter the surface wind circulation. This is known as a positive feedback loop, between the wind circulation and the sea surface temperature. There are two stable states of the feedback loop. One state is a cold east Pacific associated with strong south-east trade winds and the other state is a warm east Pacific with weaker south-east winds, the familiar El Niño.

Though we are still some way off a complete understanding of the processes which cause the ENSO, there has been considerable progress in the measurement of the phenomena. In recent years there has been the deployment of an oceanographic and meteorological observing system over the tropical Pacific Ocean. This array of instrumentation is providing real-time information on the currents, sea level, surface and thermocline temperatures, and surface winds. These observations can then be used in coupled ocean–atmosphere models to provide forecasts of the El Niño many months in advance. These new measurements not only provide the initial conditions for the models, but they will also provide verification of the models' ability to forecast El Niño, which in turn should lead to an improved understanding of the phenomena.

10.4   Long-term variations in the ocean–atmosphere system

In investigating longer-term variability of the climate system, in particular the ocean–atmosphere components and their interaction, we have a paucity of observations. The systematic collection of observations of the sea surface temperature and currents only go back 200 years, whilst deep ocean measurements of temperature and salinity go back only 100 years. The quality of the surface observations from merchant ships was often compromised by the method of sampling but, with careful scientific analysis, corrections have been possible. Higher-quality observations of the ocean temperature and salinity across the major ocean basins were taken during the International Geophysical Year (IGY) in 1957. Some of these sections have been repeated, but only at infrequent intervals of 10 or 15 years. One station at Bermuda has been monitoring the temperature and salinity in the whole water column since the 1950s, and this is the longest series of deep ocean measurements available.

Even on land, surface temperature observations extend back only 300 years at most, of which the central England time series is one of the longest, extending to 1721. The latter series is based on both temperature measurements and weather diaries to obtain a reasonably reliable record. The temperature measurements were often of low quality due to poor exposure of the thermometers. Considerable judgement and care is therefore required to obtain a consistent time series from the data. Systematic measurements of the surface meteorology over the oceans commenced in 1853, but regular daily soundings of the troposphere did not commence until the 1940s.

We will now consider what the observations have told us about the climate over this period of time. The most well-known observation is the global surface warming of the Earth observed in this century (Figure 10.9). First, this record is based on surface temperature observations (at 1 m height on land and at the sea surface). The global temperature shows a rise of about 0.5°C from 1860 to 1990. This rise in temperature, however, is far from steady over the period, with the largest increase between 1910 and 1940, followed by a small cooling for 20 years before resuming a rise in the last two decades.

Secondly, the warming signal is stronger on land than in the ocean, which may indicate a lag effect in the oceans. Thirdly, there are areas like the northern Atlantic Ocean which has shown a systematic cooling in the last 20 years, when global warming has been clearly distinguished.

Explanations of this temperature record have relied on the measured systematic rise in the atmospheric radiatively active gases (e.g. carbon dioxide and methane) over this period. Though the evidence is

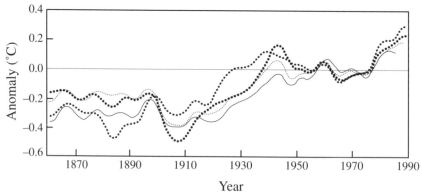

Figure 10.9   Global sea surface (solid and dashed) and land air temperature anomalies (large dots) relative to the period 1951–80. Sea surface temperatures are from Farmer *et al.* (1989) (solid) and the UK Meteorological Office (dashed)

mounting in support of this explanation, it has still not been proved beyond reasonable doubt. The proof of the association of greenhouse gases with global warming is of considerable interest because it raises philosophical questions on the nature of scientific proof in environmental problems and on actions that society can take to ameliorate the possible effects. Many scientists have looked at the evidence and found the case unproved. However, another group of scientists have found evidence from many diverse sources, including mathematical models of the system, and concluded that there is an effect. Because of the potential climate effects of unrestrained increases in greenhouse gases in the next 20 to 100 years, the Intergovernmental Panel on Climate Change have taken the pragmatic approach of publishing new evidence every five years. This should allow the international community of politicians, decision makers and the public to be appraised of the best advice on global warming from the scientists.

The instrumental record shows that there is considerable variability on a variety of time scales but, to focus the discussion, we will consider a few examples which have been well studied.

On the decadal time scale, fluctuations between the rainfall in the Sahel (a region which borders the southern fringe of the Sahara Desert in Africa) and the sea surface temperature in the tropical Atlantic Ocean have been discovered. This region suffered a number of devastating droughts in the 1970s and 1980s, whilst in the 1950s it experienced reasonably wet seasons. The difference between the drier and wetter periods is accompanied by major changes in the sea surface temperature throughout the global ocean. Subsequent studies, using global atmospheric models, have shown that the positions of the main rainfall zones in the tropics are sensitive to anomalies in sea surface temperature, which has been demonstrated for the shorter period of ENSO in Section 10.3. In the drier years of the Sahel, the west African monsoon does not penetrate as far north as in the wet years. The lack of northern penetration of the monsoon is associated with warmer than normal water in the south and equatorial Atlantic, Indian and Pacific Oceans, and cooler than normal water in the northern oceans. This example shows the global scale of this ocean–atmosphere interaction. A second example of decadal variability is that associated with one ocean, the northern North Atlantic, during the 1960s and 1970s, as shown in Figure 10.10. During this period, surface waters became

Figure 10.10   Monthly averages of salinity at eleven depths at station *Bravo* from 1964 to 1973

fresher (the salinity decreased) over a large region. The pool of fresh water, known as the Great Salinity Anomaly (GSA), moved gently around the sub-polar gyre from Greenland to Scotland in a period of over one decade. This freshwater cap prevented the deep overturning of water during periods of strong cooling in late winter. In turn, there was a reduction in the production of North Atlantic Deep Water, which has been shown to be the main water mass in the abyssal depths of 1000 to 2000 m of the world ocean and is associated with the deeper arm of the thermohaline circulation. How did the GSA develop? Some evidence suggests that it was caused by a large export of fresh water, associated with pack ice, from the Arctic Ocean into the Greenland Sea and then into the North Atlantic Ocean. Whatever the cause, the GSA indicates that small changes in the climate system, in this case fresh water, may produce profound changes on the deep ocean and its circulation.

Because the instrumental record is at best 200 years long, other methods are required to measure climate variations in the past. Scientists have developed a relatively new science of palaeoclimatology and palaeooceanography to provide this information. These sciences use a whole range of evidence from fossils, pollen, lake levels, deep-sea ocean sediments, glaciers and ice sheets to infer past variations in the climate. This is not the place to discuss all the methods and evidence that have been obtained from this science. What I will do is point to some of the results which are relevant to understanding atmosphere–ocean interactions and their role in climate change.

sOne of the notable examples is from the ice cores drilled in the Antarctic and Greenland ice sheets. These ice cores, of over 3000 m in length, contain records of past climate extending through the last glacial maximum, 18 000 years Before Present (BP), to the previous interglacial at 150 000 years BP. Small air bubbles trapped in the ice when it was first laid down as snow provide a time history of the composition of the atmosphere well before any anthropogenic influences occurred. In particular, they provide a unique trace of the natural variations in CO_2 and methane, as well as providing a set of baseline measurements before man's influence became detectable. The ice also provides other information on environmental conditions when the ice was first laid down. Oxygen isotopic composition in the ice provides a record of ocean palaeotemperature and activity levels record past volcanic events, whilst dust measurements provide an index of the storminess of the region. The length and the high temporal resolution of the records provide unique windows into past variations in climate. For example, past changes in the North Atlantic current can be inferred from the palaeo-ocean temperature in the Greenland ice cores. One core has shown that the North Atlantic varied between warm and cold conditions on the very short time scales of 20–50 years. These rapid changes in the North Atlantic Ocean, in turn, would have had a profound influence on the northern European climate. One interesting aspect of these changes in climate is that they appear to be confined to the earlier interglacial period (known as the Eemian period), whilst during the past 10 000 years (known as the Holocene), the North Atlantic temperature has been very stable. The instability of climate in the northern North Atlantic is of considerable interest to oceanographers and climate scientists as this region is known to be a major control of the ocean's thermohaline circulation.

One of the problems with the ice-core data is that it is limited to Greenland and Antarctica, and therefore some climate effects will be local to the site whilst others will be measures of global change. Since CO_2 and methane are fairly uniform throughout the atmosphere, the ice cap records provide excellent sources of global concentration. On

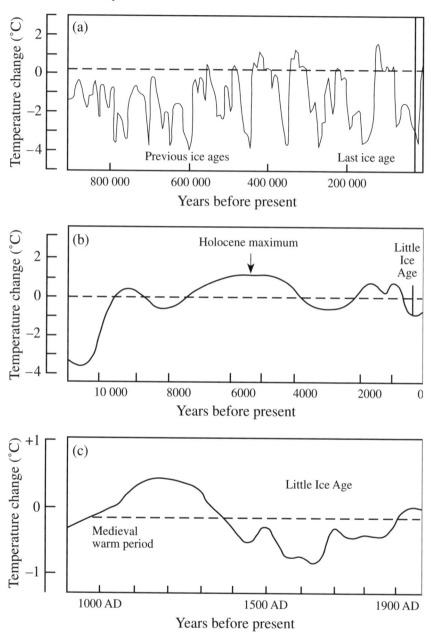

Figure 10.11 Schematic diagrams of global temperature variations on three time scales: (a) the last million years, (b) the last 10 000 years, and (c) the last 1000 years. The dashed line represents conditions near the beginning of the twentieth century

the other hand, the acidity levels in the Greenland ice sheet will be strongly affected by relatively local volcanic events in the Iceland area, but probably weakly affected by major low-latitude eruptions, such as the recent Pinatubo event in the Philippines in 1991.

The ice-core information is one source of information for the determination of past climates. Another source is the deep-ocean sediment cores, obtained from drilling programmes such as the Deep Sea Drilling Programme (DSDP) and, more recently, the Ocean Drilling Programme (ODP). These cores extend the record of climate change back millions of years. Because the changes in the deep ocean are relatively slow (cores have been obtained in all the ocean basins except the Arctic Ocean), it is possible to obtain a very reliable indication of climate change.

Cores from the Santa Barbara basin and the equatorial east Pacific have shown the annual deposition of diatoms and fish scales. The thickness of these fine or laminated sediments provides a record of productivity in the surface layers of the water column. This productivity, as we have mentioned in Section 10.3, is strongly influenced by El Niño; hence, the thickness of sediment or density of fish scales provides a unique record of El Niño events over the last 2000 years, a period well beyond the length of other records. These records indicate that El Niño is modulated by longer-period changes on time scales of 20–100 years. As yet, no mechanism for these variations has been established.

These examples show the diversity of information that has been gleaned from the ocean record on the past variations in climate. Figure 10.11 shows a reconstruction of global temperature over the past million years, inferred from palaeoclimate techniques. From this we can see that the glacial–interglacial cycles are associated with global temperature changes of about 4 to 5°C, whilst the Little Ice Age in Europe was associated with a temperature change of 1°C. These long-term fluctuations in climate allow us to view the present problems of climate change with respect to natural changes in the past. The scientific understanding from the study of past variations, coupled with the intense study of the present system, should allow better predictions of climate change to be made.

10.5   Mechanisms of climate change

From our cursory examination of the climate record (Figure 10.11), we see that there have been large changes in the Earth's climate both globally, associated with the waxing and waning of the ice sheets, and regionally, for instance, in the northern North Atlantic basin during the Little Ice Age. What are the mechanisms which give rise to these significant changes in the system?

First, we will consider the physical constraints on the time scales of the system, depicted in Figure 10.12. The climate system is composed of five basic components: the atmosphere, the ocean, the cryosphere, which is composed of ice sheets, mountain glaciers and pack ice, the land and its hydrology, and finally the biosphere. The biosphere includes all aspects of the living system on land, the atmosphere and the ocean. The latter is the most complex component and probably the least understood part of the climate system. All these components interact with other components of the system to produce climate variability on a vast range of time scales.

The atmosphere provides the shortest time scale in the system; the time for a parcel of air to travel around the globe defines the global circulation time scale, which is about 10 days in the troposphere. The atmosphere transports heat, momentum and water in all its three phases, as well as radiatively active gases (such as CO_2, methane and ozone) and particulates (such as sulphates) around the globe. The radiation time scale of the atmosphere (defined in Section 1.4) is a little longer, at about 30 days. The atmosphere is the most capricious part of the climate system. As we have seen, the clouds, which at any one time occupy about 50% of the surface area of the globe, are extremely important for the regulation of the Earth's surface temperature (see Chapter 2).

By contrast, the underlying oceans have a larger range of time scales, from months to 2000 years. The surface layers, because they are being continually driven by surface winds, solar heating and cooling by evaporation, are relatively well mixed. This mixing depth is typically about 20 m in summer to over 100 m in winter. The mixed layer has a time scale of months and, as we have seen in Section 10.2, it is associated with many short-term climate interactions with the atmosphere. By contrast, the thermocline circulations associated with

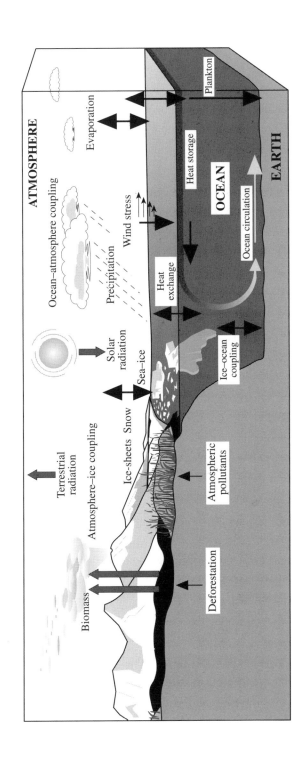

Figure 10.12 A schematic view of the Earth's climate system showing the roles of land, atmosphere, oceans and sea ice. Reproduced with permission of Mason Publishing

the major gyre systems have time scales of years to decades. The circulation time scale of the large sub-tropical gyres is about a decade. For example, the Great Salinity Anomaly in the northern Atlantic Ocean (Figure 10.10) took slightly longer than a decade to travel around the sub-polar gyre.

The thermal time scale of the gyres can be estimated from the average depth of the main thermocline. If it is assumed that the average depth is about 500 m, then this gives a time scale of about 18 years (see Section 1.4). These gyre circulations are driven by the overlying wind system by Ekman pumping, as well as the addition and removal of buoyancy at the sea surface. Therefore, they may hold the key to unravelling the decadal variability we observe in the climate system.

The deeper abyssal oceans provide the longest time scale of the ocean. The driving force for the abyssal oceans is focused on the high-latitude regions of the globe, wherein the abyssal water masses are formed. The circulation time scale is of the order of centuries to a millennium. For example, the North Atlantic creates about $20 \times 10^6 \, \text{m}^3 \, \text{s}^{-1}$ of North Atlantic Deep Water (NADW). If we assume that the abyssal ocean is composed entirely of NADW, and take the volume of the abyssal ocean to be about $1.0 \times 10^{18} \, \text{m}^3$, then the time scale for the deep overturning circulation is 1585 years. The Atlantic Ocean has a quarter of the total volume; therefore, the circulation time is about 400 years.

The third component is the cryosphere, wherein about 2% of the Earth's total water and 75% of the Earth's fresh water is stored. One of the shortest time scales is that of the surface pack ice which floats on the Arctic Ocean and around the Antarctic continent, and which has a thickness of a few metres. Its growth and decay is controlled by heat fluxes and precipitation between the surface and the atmosphere, and by the ocean heat fluxes from below the ice. Because of these factors, sea ice exhibits considerable growth and decay on seasonal and interannual time scales. Furthermore, because it is driven by surface winds and ocean currents, it can transfer large quantities of fresh water from its place of formation to its place of melting, and these will need to be replaced by the ocean and atmosphere circulation. An example of this process occurs in the Antarctic where the pack ice is driven northwards from the coast by intense offshore winds, thus providing open water adjacent to the coast. This in turn allows more ice to be frozen at the coast by intense cooling of the ocean by the Antarctic winds. Pack ice is generally low in salinity due to brine removal back into the ocean during the freezing process. Hence, the ocean becomes more saline beneath the ice. The combination of high salinity and low temperatures of these waters leads to the production of the most dense water mass in the ocean: Antarctic Bottom Water.

Sea ice is also important as a regulator of ocean temperature due to its low thermal conductivity, and hence heat transfer from the ocean to the atmosphere is considerably reduced in ice-covered regions of the Arctic and Antarctic Oceans.

The mountain glaciers and ice sheets/ice shelves provide the longer time scales of the cryosphere. The growth and decay of mountain glaciers occur on time scales of decades to centuries, whilst the major ice sheets have time scales of many thousands of years. A dramatic example of the decay of glaciers is that seen in the last 150 years by most of the world's glaciers associated with the global rise in surface temperatures. In general, glaciers are in quasi-balance between the growth of ice at the head of the glacier, mainly due to snowfall accumulation over many years, and ablation processes at the lower end of the glacier. These ablation processes include surface melting and evaporation, melting at the base of the major ice sheets and calving when the ice breaks off into the sea. The growth of a glacier is therefore dependent not only on the temperature, but also on the rate of snowfall at the head. In addition, because of the vast accumulations of ice in Greenland and Antarctica (of the order of 2 km thick), there is lowering of the Earth's lithosphere and asthenosphere by isostatic adjustment. This isostatic adjustment is seen today in Scandinavia, where the land is rising at the rate of 1 cm per year following the melting of the ice sheet over 10 000 years ago. This adjustment process brings an additional longer time scale into the dynamics of ice sheets.

The Milankovitch theory for glacial cycles implicitly relies on feedback within the climate system to amplify the relatively small changes in seasonal solar radiation incident on the atmosphere to produce large changes in the ice sheets. Milankovitch placed his emphasis on the crucial role of solar radiation in the summer-time at 60°N. When the solar radiation due to orbital effects was diminished, there would be decreased ablation of the ice in summer, and therefore the ice sheet would grow slowly from one year to the next. As the area covered by the ice sheet increased, the albedo would increase, thus reducing the amount of solar radiation absorbed by the ice and further decreasing the amount of ablation. The theory does not take into account many of the physical processes which influence the growth of an ice sheet; for example, the changes in snowfall due to the increased level of the ice sheet, and the influence of clouds and aerosols on the surface radiation budget of the ice sheet. Despite the theory being very simple, it has been widely accepted by the scientific community, because the analyses of deep-sea cores have shown very clearly that the main frequencies of variability are associated with these forcing frequencies of the solar radiation.

The land surface and its associated hydrology is the fourth component of the climate system. The air temperature above the surface is controlled by the energy budget over that surface, discussed in Chapter 5. The absorption of solar radiation will depend on the albedo of the surface, whether it is snow, ice, sand desert, or its type and cover of vegetation (Table 1.3), whilst the heat loss by evaporation will depend on the availability of water stored in the vegetation or soil, in lakes and rivers, or as groundwater. The albedo is an exceptionally important variable in the climate system. The spring melt of snow cover over the continents is an example of the influence of albedo. Whilst the snow cover exists, only a relatively small proportion of the solar radiation is absorbed by the surface. At the same time, heat is lost from the surface by long-wave radiation and sublimation to the atmosphere, and therefore little or no heat is available for melting. The heat balance maintains the snow-covered ground. Now, if the snow is removed from a soil surface, the albedo will decrease from about 80% to 20% and, as a consequence, the absorbed solar radiation will increase by a factor of 4 and the surface temperature will rise. To balance the absorbed solar heat, the heat loss from the surface would increase by a similar amount, and thus the air above the surface would also be warmed rapidly. Over the continents in the spring-time, the increasing solar radiation provides the stimulus for the rapid thaw of the snow cover, which may take place over a period as short as a few days.

Albedo changes will also occur where there are major changes in land use, for instance, associated with major deforestation or with the withering of vegetation associated with drought. There is some evidence that the prolonged droughts in the Sahel region in the 1970s and 1980s involved albedo feedback. In this case the drought led to an increase in surface albedo, which in turn decreased the propensity for rainfall in this region.

The final component of the climate system is the biosphere. This component is fundamentally important, as it has been part of the process in developing the present-day composition of the atmosphere. The GAIA hypothesis takes this a stage further, by suggesting that the biosphere has regulated its own climate from early in Earth's history by changing the composition of the atmosphere.

We have already seen how land use and its associated vegetation provide an important example of albedo feedback. A second example of this albedo feedback has been hypothesized in the ocean, in connection with coccolithophores. These phytoplankton, because of their calcite shells, produce a chalky appearance to the water when they bloom in northern North Atlantic waters. The albedo of sea water is thus increased by the appearance of the bloom, which results in reduction of the absorbed solar radiation and a cooling of the surface waters.

Vegetation is not only important in albedo feedback. It also has the capacity, with the underlying soil and hydrology, to control the energy budget of the surface and of the air above it. The tropical rainforest is an example of an ecosystem where the water and nutrients are recycled very efficiently. When the forest is removed, the water is evaporated very rapidly and, because of the poor soils, the area may revert quickly to low-quality scrubland.

In this section we have seen that, generally, there are five components which interact to produce our climate. Sometimes the interaction may involve only two major components, for example, the ENSO, which is an interaction between the tropical ocean and atmosphere. It may involve three or four components, such as the Sahel decadal droughts, which involve the land and its surface hydrology, the biosphere, the ocean surface and the atmosphere.

The inherent complexity of the system indicates that systematic study of climate variability and change requires new methods to enable us to make this subject into a quantitative science from which predictions will be made.

The mathematical climate model provides the most powerful tool for this study. The basis of the model is a set of equations which describe the processes occurring in each of the components of the system, together with equations which describe the transfer of quantities between each component of the system; for example, the transfer of momentum or the exchange of carbon dioxide between the atmosphere and ocean. In Chapter 9 we showed some examples of ocean, atmosphere and land models which are being used to study aspects of the climate problem. In addition, ice-sheet models have been developed and are being actively validated from the known record. Though it would be foolish to suggest that these models are perfect, they do have sufficient authenticity to be used for a climate model of the physical system. At present, perhaps the least understood component is the biosphere, particularly in the ocean. The complexity of the marine biology component, and the spatial and temporal sampling problem associated with ships and satellites, makes it difficult to measure the rapid changes associated with plankton blooms over the globe. This means that, at present, a complete climate system model which has all five components interacting with each other is some way off being realized. For the present, we may have to use more limited climate models with perhaps three components of the system to understand climate change and variability.

Estimates of global warming by greenhouse gases have been made without the use of ice-sheet models or ocean biology. We are able to do this because the actual atmosphere concentrations are well determined for the last 45 years and, from this record, estimates of greenhouse gas concentrations in the next 50 years are obtained. These concentrations can be put directly into coupled ocean–atmosphere models, from which predictions of surface temperature on a global and regional scale can be made.

However, these models cannot address the question of how the anthropogenic carbon dioxide is repartitioned between the atmosphere, ocean, land and biosphere. At present, we know that only half the anthropogenic component remains in the atmosphere, but whether the missing carbon is taken up by the land biosphere or by the ocean we do not know. To address these problems we will need a complete global carbon model, which will certainly include the ocean, atmosphere, land and biosphere, and possibly the cryosphere. Such a model will have to be verified using global data sets for all the major components of the mobile carbon system.

10.6   Perspectives in ocean and atmospheric science

Both meteorology and oceanography are observational sciences, and therefore the growth in understanding of how these complex systems work has been largely dependent on improvements in the methods of measurement. For example, the establishment of the synoptic network of upper air stations using the radiosonde in the 1940s and 1950s provided, for the first time, daily information on the pressure and wind fields in the troposphere and lower stratosphere. These observations led to an improved understanding of the development of weather systems which, in turn, laid the foundations for numerical weather predictions. Similarly in oceanography, it has been seen that the neutrally buoyant float provided the first important evidence for mesoscale eddies in the deep ocean, as well as some of the first observations of the abyssal western boundary currents. Both of these phenomena had not been previously observed because of the coarse horizontal resolution of hydrographic data.

Despite the considerable improvements in instrumentation, both meteorologists and oceanographers today are confronted with questions which, if they are to be answered satisfactorily, require additional methods of measurement to those discussed previously. One important question involves the role of the ocean in seasonal and interannual variations in climate.

In Section 10.3 it was shown that the ENSO is the most prominent signal in the atmosphere–ocean system on interannual time scales. To be able to provide forecasts of ENSO, tropical ocean models are coupled to global models of the atmosphere. These models, however, require regular measurements of the atmosphere, and of the tropical oceans, to provide the initial conditions from which the forecasts can be made. Fortunately, we have a reasonable measurement system for the troposphere and lower stratosphere which provides the initial conditions for weather forecasts across the globe. For the ocean, a new system called the Tropical Pacific Ocean Observing System has started to provide regular observations from Papua New Guinea to the Americas. This system involves moored and free-floating buoys, volunteer observing ships and a network of tide gauges. The ocean observations include surface wind, heat and freshwater fluxes, sea surface temperature and sea level, as well as the vertical profiles of temperature, salinity and currents down to the depth of the equatorial thermocline. The data is relayed from the instruments by satellite to climate centres around the world to provide the data required by the coupled models for ENSO predictions. This is the first step in the provision of regular and high-quality measurements of the surface layers of the ocean.

The Tropical Pacific Ocean Observing System illustrates the requirement to provide a methodological approach to measurements if we are to use all our knowledge to address the important questions discussed in Section 10.5. Scientists are currently designing a global climate observing system, which will have an ocean component, called the global ocean observing system. The scientific understanding gained from the recent and present international research programmes, such as the World Ocean Circulation Experiment and the Tropical Ocean Global Atmosphere (TOGA) experiments, will be used to design a system from which climate forecasts can be made.

An extremely important part of a global observation system is the use of space platforms to observe the Earth. The development of space instruments to measure the properties of the atmosphere, and the underlying ocean and land surfaces, has been very impressive in recent decades. Weather prediction has been greatly enhanced by the use of radiometers which can measure the reflective energy in the visible or the emitted radiation in the thermal infra-red bands from clouds. This enables them to pinpoint tropical and extra-tropical cyclones, and frontal cloud bands, in remote regions of the world.

First, we will discuss the satellite orbits and then the measurements which can be made. A satellite can be placed into a geostationary orbit over the equator or it can be placed in a polar orbit. The geostationary satellite has a nadir view over the same position on the equator, but is limited to a maximum latitude of 60°. For example, a set of five meteorological satellites can provide continuous monitoring over most of the globe. A disadvantage is the height of the satellite above the Earth's surface (about 36 000 km), and this therefore limits the horizontal resolution of the radiometers. In contrast, the polar orbiting satellite can provide an improved global coverage and with a lower orbit of between 800 and 1200 km above the Earth's surface. The improved spatial resolution is bought at the expense of the temporal resolution. For example, the GEOSAT satellite, which measured sea level, covers the same part of the globe only once every 17 days.

The selection of the oceanographic and meteorological parameters which can be determined from the present generation of instruments is shown in Table 10.1. The instruments measure either (i) the passive emission of radiation from a wavelength between the infra-red and microwave, or the reflection of sunlight, or (ii) uses a radar or lidar on the satellite. The measurement of either atmospheric temperature, cloud-top temperature, or the sea surface or land surface temperature relies on the emission of thermal infra-red radiation. By using multi-spectral sensors, more discrimination in either the vertical structure of the atmosphere or in the surface properties can be obtained.

Until recently, most meteorological satellite systems were passive. They measured the radiation emitted in the infra-red and microwave regions of the spectrum, and they also monitored reflected solar radiation, as shown in Table 10.1. Passive sensors in the visible are able to delineate regions of clouds, ice boundaries, and the colour of the land and sea surfaces during daylight hours. However, thermal infra-red sensors, which measure the radiation temperature of the Earth's surface or cloud tops, are able to delineate variations both during the day and the night. The thermal emission from the sea surface can be used to obtain the sea surface temperature using known values of the emissivity. This sea surface temperature is representative of the upper 100 μm, and it is therefore different from the bulk temperature typical of the upper 1–3 m measured by ships. However, except under low wind conditions, the differences are usually less than 1ºC. Unfortunately for sea surface temperature measurements, the atmosphere also emits radiation at infra-red wavelengths due to the presence of carbon dioxide, ozone and water vapour, and therefore the signals have to be corrected for atmospheric effects. Furthermore, clouds will absorb all thermal radiation emitted from the sea surface, and therefore the technique is limited to cloud-free regions. Microwave energy is also emitted from the sea surface and, though much less than the emitted infra-red energy, at certain wavelengths it can be transmitted through non-precipitating clouds and used to obtain sea surface temperature measurements. The thermal emission of radiation by the atmosphere has been used to obtain the vertical distribution of temperature in the atmosphere. The principle of the instrument used is that different wavelengths of infra-red and microwave radiation originate at different levels in the atmosphere. The radiance at each wavelength can therefore be related to the temperature at the appropriate level. The

Table 10.1 Selection of atmospheric and oceanic parameters measured by satellite

Environmental parameter	Satellite sensor	Accuracy	Method
Tropospheric wind	Visible/infra-red	$\pm 4\,\mathrm{m\,s^{-1}}; \pm 20^{\circ}$	Cloud displacement from geostationary satellite
Surface winds	Scatterometer (radar)	$\pm 2\,\mathrm{m\,s^{-1}}; \pm 15^{\circ}$	Scattering of radar beam by sea surface
Waves	Altimeter (radar)	$\pm 0.3\,\mathrm{m}$	Shape of reflected pulse related to significant wave height
Sea surface elevation	Altimeter (radar)	$\pm 0.02\,\mathrm{m}$	Time delay of radar pulse
Vertical atmospheric temperature profile	Infra-red/passive microwave	$\pm 2\,\mathrm{K}$	Multichannel radiometer samples emitted thermal radiation from different levels of atmosphere
Sea surface temperature	Infra-red/passive microwave	$\pm 0.3\,\mathrm{K}$	Thermal radiation from sea surface Corrections for atmospheric contamination Infra-red limited to cloud-free areas
Solar radiation at sea surface	Visible/near infra-red	$\pm 10\,\mathrm{W\,m^{-2}}$	Measures cloud-top temperature
Precipitation	Visible/infra-red	Qualitative	Empirically related to precipitation
	Passive microwave	100%	Measures liquid water content of atmospheric column
Ice cover	Visible/infra-red	—	High resolution in visible and infra-red
	Passive microwave		Low resolution, but all weather, in microwave

vertical resolution is not as good as that obtained from a radiosonde, being about 5 km, but the horizontal resolution is much better, being typically 100 km for infra-red observation and 250 km for passive microwave measurements.

Passive microwave sensors have been used to determine the total water vapour and the total liquid water content of the atmospheric column from simultaneous measurements of the water-vapour absorption line at 1.36 cm wavelength and the liquid water absorption band at 0.97 cm wavelength. From such measurements, global maps of water vapour and liquid water content have been produced. These maps would have been unobtainable with the same resolution from conventional observations. Absorption of microwave radiation at a wavelength of 1.55 cm is dependent on the number and size of the water droplets in the atmospheric column, and this fact has been used to deduce precipitation rates. It is interesting that this technique is most suited to ocean areas where the emission of contaminating microwave radiation from the ocean surface is low compared with the emission from land surfaces. These variations in surface emissivity in the microwave spectrum have been used to detect the distribution of sea ice. The higher microwave emissivity of sea ice compared with sea water allows identification of the boundary between sea ice and open water even in cloudy conditions. Weekly variations in sea-ice cover can thus be monitored by the satellite network. In addition, it is known that the emissivity of ice more than one year old is lower than that of newly-formed ice, and therefore the age distribution of the ice can be mapped.

In the field of oceanography, the development of active microwave satellite instruments is seen as the most promising avenue for obtaining quantitative information about the ocean surface. In 1978, the first oceanographic satellite, SEASAT, was launched. These satellites carry three different types of radar. The first radar, known as the scatterometer, measures the back-scattered radiation from the sea surface. The intensity of the back-scattered radiation is proportional to the surface-wind stress, whilst the direction of the scattering is related to the direction of the surface stress. Although the theory of the scattering at this wavelength is not well understood, the technique can provide accurate wind measurements over the world's oceans. Second, a synthetic aperture radar, SAR, provides radar images of the sea surface. These images give information about the surface wave spectrum and the direction of the waves. Surprisingly, the radar images also detect internal waves in the ocean, as well as variations in bottom topography in water which is many tens of metres deep. Since SAR wavelengths are absorbed in the upper centimetre of the ocean, this effect is an indirect influence of bottom topography on the surface wave field. The third radar is a high-precision radar altimeter which is able to measure the distance between the ocean surface and the satellite to a precision of ± 2 cm. Variations in the geometric distance between the sea surface and satellite are related to variations in the satellite orbit, variations in the surface geoid and the ocean surface slope associated with ocean currents. The first effect can be eliminated by accurate position-fixing systems, but distinguishing geoid variations from dynamical sea-level variations is very difficult unless independent measurements of the geoid or sea level are available. However, since the geoid is static, repeated passes of the altimeter over a given area can distinguish time-dependent changes in the sea surface elevation. These satellites are able to detect time variations of mesoscale eddies in the world ocean, as shown in Figure 7.9, and it is now feasible to obtain eddy maps of the ocean similar to the now familiar atmospheric weather maps. In the equatorial ocean, sea-level variations are associated with displacement in the shallow equatorial thermocline. High sea levels are generally associated with a deep thermocline and vice versa. Thus, it is possible that altimeter measurements may be used to obtain routine observations of the thermocline displacement, as well as the heat content in the upper ocean. This technique would be invaluable for measuring seasonal and interannual variations in the equatorial ocean associated with ENSO and other large-scale phenomena. With the improved precision of the altimeter, to ± 2 cm, and with high-quality determinations of the surface geoid, it is now feasible to determine surface geostrophic currents from the satellite measurements, since:

$$v_g(0) = \frac{g}{f}\frac{\partial \xi}{\partial x} + \frac{1}{f\rho_0}\frac{\partial p_0}{\partial x} \qquad (10.1)$$

where $v_g(0)$ is the surface geostrophic current, ξ is the height of the sea surface and p_0 is surface pressure. All other terms have their conventional meaning.

From meteorological surface pressure measurements, the second term in equation 10.1 can be determined, and hence v_g at the surface is obtained (Figure 10.13). Moreover, from the scatterometer wind-stress data, the wind-drift current, v_e, can be calculated. Thus, the total surface current $v_g + v_e$ could be determined.

We have seen that the satellites can measure sea surface temperature and sea level over the globe. However, because of the lack of transparency of sea water to electromagnetic radiation, the satellites are limited in what they can tell us about the deep ocean. For the interior

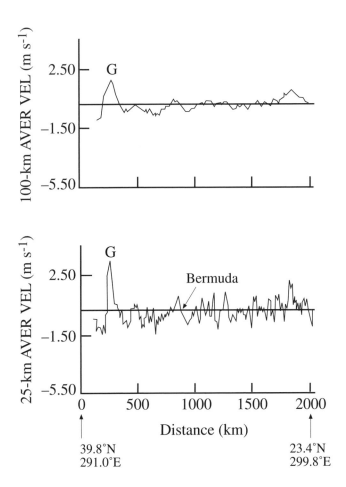

Figure 10.13   Geostrophic velocity determined by satellite altimetry along a track crossing the Gulf Stream and Bermuda. The Gulf Stream (G) and variations due to eddies east of the Gulf Stream can be seen. Two different averaging intervals are displayed

ocean, new technologies and measurement methods are being developed. The relative ease with which sound can travel through the oceans is being used to measure global ocean temperature. We can recall from Chapter 2 that the velocity of sound in sea water is strongly influenced by temperature and pressure. These two properties conspire to produce a minimum velocity at depths of about 1000 m, which is known as the sound or SOFAR channel. Low-frequency sound (70 Hz) can travel in the sound channel for thousands of kilometres with little attenuation. An experiment in 1991 transmitted sound from Heard Island in the Southern Ocean to receivers as far away as the Pacific and Atlantic coasts of North America, a distance of 16 000 km. The travel time is dependent on the temperature of the ocean through which the sound travels; hence, changes in travel time from one year to another will be a measure of the average temperature change between the source and receivers. This Acoustic Thermometry of Ocean Climate (ATOC) is being tested in the North Pacific Ocean, with sound sources at Hawaii and California, and receivers around the Pacific rim. This array will be used to test whether thermal anomalies that have been seen from repeated hydrographic stations can be detected and their evolution monitored.

Because of the costs of measurements from ships and the inherent problems in sampling of the ocean, both spatially and temporally, there is considerable interest in the use of autonomous instruments. The recent development of the Autonomous LAgrangian Circulation Explorer (ALACE) float is one such example (see Chapter 6). A large deployment of these instruments in the Pacific and Southern Oceans during the WOCE experiment is now providing a unique view of the large-scale structure of ocean currents.

Another development is the Autosub. The objective of this project is to provide autonomous hydrographic sections across ocean basins. The key elements of the Autosub are a very low drag body, which has an efficient propulsion system, together with sensors that retain their accuracy over many months. In similarity with ALACE floats, it will surface to telemeter its position and measurements to a satellite.

Though the measurements of temperature, salinity and pressure are the main objectives of these autonomous vehicles and floats, there are also biological and chemical sensors which can be attached. These sensors can measure the optical properties of the water to determine chlorophyll and other pigments derived from phytoplankton, and chemical properties such as dissolved carbon dioxide, oxygen and nitrates.

These developments in instrumentation will not on their own provide the answers to the many questions being asked of ocean scientists, meteorologists and climatologists, because of the complexity of the system. To be able to discern the global and regional changes in the ocean, we require mathematical models of the world ocean into which these observations may be added. The model will predict the variables such as currents, temperature and salinity, density and pressure at a lattice of grid points, as discussed in Chapter 9. These predictions can be considerably improved upon by the addition of observations. The assimilation process allows observations to be smoothly introduced into the model as it is integrated forward in time. The data sets obtained are dynamically consistent, and this provides an improvement on both the original observations and the model.

These global ocean models are under considerable development to improve their ability to handle the complex topography of the ocean basins and coastlines, to transport and mix properties within the ocean, and to produce improved transports of heat and fresh water between ocean basins. This development goes hand-in-hand with the expansion of data collection. At the end of the WOCE experiment (1990–97), more measurements will have been taken than in all the previous oceanographic experiments.

These models will be run on parallel computer systems to obtain the necessary computational power to run them for many decades at a resolution sufficient to resolve the most energetic turbulent eddies in the ocean. They will be used to predict climate changes years, if not decades, ahead.

We should not neglect the problems of how these ocean models and atmospheric models are coupled together. The interchange of properties between the atmosphere and ocean occurs in turbulent boundary layers (Chapter 7). These processes at present are not well understood, though measurements have considerably improved with the advent of oceanographic satellites. These measurements of surface stress, surface waves and sea surface temperature provide useful information which can help to validate models of the ocean boundary layer.

The modelling of pack ice remains a difficult problem, residing as it does between the atmospheric and ocean boundary layers. The ice interface is far from uniform in thickness or extent. Improvements in the modelling of sea ice will require better understanding of the processes of the atmospheric boundary over the ice. The modelling of sea ice has to include the heterogeneity of ice from the thin ice at the edge of the pack to the solid pack with its associated pressure ridges and ice rafting.

It is not difficult to appreciate the considerable scope there is to improve our knowledge and understanding of the ocean–atmosphere system.

Though it is difficult to predict the particular development of a science as complex and interactive as the ocean and the atmosphere, we can see a common thread in the study. The understanding comes from a combination of observations and scientific principles which provide the basis of the mathematical models. The present challenges of the subject will mean that it continues to be an exciting and demanding science.

Glossary

adiabatic process

An adiabatic process is where a fluid (gas or liquid) changes temperature, as a result of either a change in pressure or volume of the fluid, subject to *no* heat being removed or added to the fluid from its surroundings.

The *dry adiabatic lapse rate* is the change in temperature with height of a dry air parcel caused solely by the change in atmospheric pressure. There is no heat added to or removed from the air parcel.

The *saturated adiabatic lapse rate* is the change in temperature with height of a saturated air parcel caused solely by the change in atmospheric pressure. The latent heat of condensation provides an additional heat source, and therefore it is not strictly an adiabatic process. It is often referred to as a pseudo-adiabatic process.

age of a water mass

The age of a water mass is the time elapsed since the water mass was formed by contact with the atmosphere. This can be determined by *in situ* measurements of dissolved gases, such as the family of freons whose concentrations in the atmosphere are well determined over the last 70 years.

albedo

The *planetary albedo* is the ratio of the reflected radiation from the Earth's atmosphere and surface to the incident solar radiation above the atmosphere.

The *surface albedo* is the ratio of the reflected radiation from the surface to the incident solar radiation, both diffuse and direct. It is measured across a horizontal plane above the surface.

atmospheric aerosol

A general term which refers to all particulates and droplets suspended in the atmosphere.

atmospheric boundary layer

The region of the lower atmosphere which is influenced directly by the Earth's surface. The winds in the boundary layer are affected by the frictional forces acting across the surface, whilst the thermal and moisture properties of the air will be influenced by heat and evaporative fluxes from or across the surface.

available potential energy

This is the potential energy of the atmosphere and ocean which can be converted into the kinetic energy of the circulation. The available potential energy is dependent on horizontal variations of density (ocean) and potential temperature (atmosphere). The available potential energy is considerably smaller than the potential energy.

black body

A perfect emitter and absorber of radiation at all wavelengths is known as a black body.

circulation time scale

This is the time for a fluid parcel to circulate around an ocean basin. For example, a parcel of water will take about five to ten years to travel around a sub-tropical gyre in the North Atlantic.

correlation coefficient

A statistical measure of the correspondence between two variables.

cryosphere

A generic term for the frozen part of the Earth's system in all its forms on or at the Earth's surface. This includes mountain glaciers, ice sheets, sea ice and lying snow.

diabatic process

This refers to a process which transfers heat to a fluid parcel from its surroundings. For instance, radiation is a diabatic process because it can warm or cool a fluid parcel by the absorption or emission of radiation.

Earth's orbit around the sun

The *eccentricity* is a measure of the ellipticity of the Earth's orbit around the Sun.

The *obliquity* is the tilt of the Earth with respect to the vertical through the orbital plane of the Earth around the Sun.

The *longitude* of the perihelion is the time of the year when the Earth is closest to the Sun. It is defined with respect to the Northern Hemisphere spring or vernal equinox. For example, $0°$ is the vernal equinox, $90°$ is the summer solstice, $180°$ is the autumn equinox and $270°$ is the winter solstice.

emissivity

This is a measure of the efficiency of a body for emitting radiation. A black body is a perfect emitter of radiation, and therefore has an emissivity of one.

entrainment

This is a process whereby a turbulent layer of fluid grows at the expense of a quiescent layer of fluid. For example, turbulence at the ocean surface, generated by wind and waves, will mix with quiescent fluid below, which will increase the depth of the surface turbulent layer.

equipotential surface

This is an imaginary surface around the Earth which has a constant gravitational potential energy. The acceleration of gravity due to the Earth is always normal to the equipotential surface.

frequency spectrum or power spectrum

This is a decomposition of the energy (or variance) of a fluctuating quantity into either its frequency (or period) or its wavenumber (or wavelength).

geoid

The geoid is an equipotential surface which coincides with mean sea level over a stationary ocean. Ocean currents and tides cause deviations of sea level from the geoid.

geostationary

A geostationary satellite is one which is above the equator and rotates with the same velocity as the earth's surface under the satellite. The satellite will view the same disc of the earth at all times.

halocline

A maximum in the vertical gradient of salinity. It occurs where fresh water either from rivers or meltwater from ice overlies saline ocean water. The fresh water is less dense than the ocean water, and therefore forms a vertical stable layer.

Indonesian maritime continent

This is the term used by meteorologists to describe the Indonesian region composed of islands and sea. It acts as a huge source of heat and water vapour, and powers the atmospheric circulation in the Indo-Pacific region.

inverse analysis

This is a mathematical procedure used for calculating the ocean circulation from measured properties in the ocean. These properties may include both the normal hydrographic measurements of temperature and salinity, as well as measurements of conservative chemical and radioactive tracers.

isopycnals

These are quasi-horizontal surfaces of constant density in the ocean.

Water parcels tend to mix along these surfaces.

isostasy

A theoretical balance of all large portions of the Earth's crust floating on a denser underlying layer at a depth of about 110 km.

ITCZ

The Inter-Tropical Convergence Zone (ITCZ) is a region close to the equator, where the moist trade wind systems from both hemispheres converge. The resulting uplift of air, causes very heavy rainfall at the convergence zone. The zone tends to move across the equator with the sun.

lithosphere

The rigid outer layer of the Earth which comprises the Earth's crust and the solid upper part of the mantle.

MODE

The acronym given to the Mid Ocean Dynamics Experiment (MODE), an experiment to measure meso-scale ocean eddies in a 200 km square of the North Atlantic Ocean, near Bermuda, in the 1970s. It was the largest physical oceanography experiment of its time and was supported by the USA.

ocean mixed layer

This is the term used to describe the surface layer of the ocean, which is continuously mixed by turbulent eddies generated by wind, waves and surface cooling. It is characterised by vertically uniform properties of temperature and salinity, and may extend to depths in excess of 100 m in winter.

potential temperature

A conservative property of an incompressible adiabatic fluid. In the atmosphere this property is useful as a conservative tracer of motion in the troposphere and stratosphere on a time scale of a few days. It can also be used as a tracer of ocean circulation.

pycnocline

A maximum in the vertical gradient of density which may be caused by either temperature or salinity gradients, or both factors. The main pycnocline in the ocean closely follows the main thermocline.

Rossby waves

These are quasi-horizontal planetary waves whose behaviour is determined by the northward gradient of the Coriolis parameter. They are named after the Swedish meteorologist Carl Gustav Rossby.

SOFAR channel

The SOFAR (SOund Fixing And Ranging) channel is a waveguide for sound propagation in the ocean with its axis corresponding to the minimum in sound velocity. Low-frequency sound can travel many thousands of kilometres in the SOFAR channel.

supersaturation

If the water vapour pressure is e , the supersaturation (in percent) with respect to liquid water is $(e/e_s-1)100$, where e_s is the saturation vapour pressure over the plane surface of liquid water. A supersaturation with respect to ice is defined in a similar way.

synoptic

In meteorology this refers to a characteristic horizontal scale of motion associated with extra-tropical cyclones, typically $\sim$ 1000 km to 5000 km

thermal equator

This the zone of maximum sea surface temperature which generally moves with the sun across the equator. The ITCZ is often located on the thermal equator.

thermocline

A maximum in the vertical gradient of ocean temperature. The best example is the main thermocline which is the boundary between the warm thermocline waters and the cold abyssal waters. The warm layers are less dense than the deeper and colder layer, and hence a vertical stable layer is formed.

upwelling

Upwelling occurs where cooler thermocline waters are brought to the surface to replace the warmer surface waters. It is mainly observed in tropical regions of the ocean, where cooling of several degrees may occur. The upwelling is often caused by wind events along a coastal region, along the equator, or by a tropical cyclone. These upwelled waters are usually regions of high biological productivity.

vertical stability

When lighter fluid overlies heavier fluid, the vertical stability of the system is hydrostatically stable. In contrast, when heavier fluid overlies lighter fluid, the system is unstable and there is rapid vertical mixing.

General Reading

Deacon, M.B., 1971, *Scientists and the Sea – a Study of Marine Science, 1650–1900* (London: Academic Press), 445 pp.

Deacon, M.B. (ed.), 1978, *Oceanography: Concepts and History*, Benchmark Papers in Geology Vol. 35 (London: Dowden, Hutchinson & Ross Inc.), 394 pp.

Defant, A., 1961, *Physical Oceanography*, Vols 1 (729 pp.) and 2 (598 pp.) (Oxford: Pergamon).

Dietrich, G., Kalle, K., Krauss, W. and Siedler, G., 1980, *General Oceanography: An Introduction*, 2nd edition (New York: Wiley), 626 pp.

Gill, A.E., 1982, *Atmosphere–Ocean Dynamics* (London: Academic Press), 662 pp.

Gross, G.M. and Gross, E.R. 1996, *Oceanography – A View of the Earth*, 7th edition (Englewood Cliffs, NJ: Prentice-Hall), 472 pp.

Harvey, J.G., 1976, *Atmosphere and Ocean: Our Fluid Environment* (Artemis Press), 143 pp.

Hess, S.L., 1959, *Introduction to Theoretical Meteorology* (New York: Holt Rinehart & Winston), 362 pp.

Hill, M.N., 1963, *The Sea: Ideas and Observations, Vol. 1. Physical Oceanography* (New York: Wiley-Interscience), 864 pp.

Holton, J.R., 1979, *An Introduction to Dynamical Meteorology*, 2nd edition (New York: Academic Press), 391 pp.

McLellan, H.J., 1965, *Elements of Physical Oceanography* (Oxford: Pergamon), 151 pp.

Maury, M.F., 1853, *The Physical Geography of the Sea* (London: Sampson Low & Son), 274 pp.

Monin, A.S., 1972, *Weather Forecasting as a Problem in Physics* (Cambridge, MA: MIT Press), 199 pp.

Monin, A.S., 1986, *Introduction to the Theory of Climate* (Dordrecht: Reidel), 261 pp.

Neumann, G. and Pierson, W., 1966, *Principles of Physical Oceanography* (Englewood Cliffs, NJ: Prentice-Hall), 545 pp.

Perry, A.H. and Walker, J.M., 1977, *The Ocean–Atmosphere System* (London: Longman), 160 pp.

Pickard, G.L. and Emery, W.J., 1982, *Descriptive Physical Oceanography: an introduction*, 5th edition (Oxford: Pergamon), 320 pp.
Pond, S. and Pickard, G.L., 1983, *An Introduction to Dynamical Oceanography*, 2nd edition (Oxford: Pergamon), 329 pp.
Riehl, H., 1979, *Climate and Weather in the Tropics* (London: Academic Press), 611 pp.
Robinson, I.S., 1985, *Satellite Oceanography*, (Chichester: John Wiley), 455pp.
Summerhayes, C.P. and Thorpe, S.A., 1996, *Oceanography – An Illustrated Guide*, (Manson Publishing), 352pp.
Sverdrup, H.U., 1945, *Oceanography for Meteorologists* (London: George Allen & Unwin), 246 pp.
Sverdrup, H.U., Johnson, M.W. and Fleming, R.H., 1946, *The Oceans: Their Physics, Chemistry and General Biology* (Englewood Cliffs, NJ: Prentice-Hall), 1087 pp.
Tchernia, P., 1980, *Descriptive Regional Oceanography*, Pergamon Marine Series, Vol. 3 (Oxford: Pergamon), 253 pp.
Tolmazin, D., 1985, *Elements of Dynamic Oceanography* (London: Allen & Unwin), 181 pp.
Wallace, J.M. and Hobbs, P.V., 1977, *Atmospheric Science* (London: Academic Press), 467 pp.
Warren, B.A. and Wunsch, C., 1981, *Evolution of Physical Oceanography* (Cambridge, MA: MIT Press), 623 pp.

Further Reading and References

Chapter 1

Allen, C.W., 1958, Simons Memorial Lecture 'Solar Radiation', *Quarterly Journal of the Royal Meteorological Society*, **84**, 307–318.

Goody, R.M. and Walker, J.C.G., 1972, *Atmospheres* (Englewood Cliffs, NJ: Prentice-Hall).

Henderson-Sellers, A. and Wilson, M.F., 1983, Albedo observations in climatic research, *Philosophical Transactions of the Royal Society of London*, **A309**, 285–294.

Ramanathan, V., Barkstrom, B.R. and Harrison, E.F., 1989, Climate and the earth's radiation budget, *Physics Today*, **42**, 22–32.

Chapter 2

Allegre, C.J. and Schneider, S.H., 1994, Evolution of the earth, *Scientific American*, **271**(4), 44–51.

Brancazio, P.J. and Cameron, A.G.W., 1964, *The Origin and Evolution of Atmospheres and Ocean* (New York: Wiley), 314 pp.

Dittmar, W., 1884, Report on research into the composition of ocean water collected by H.M.S. Challenger during the years 1873–76, in *Report on the Scientific Results of the Voyage of H.M.S. Challenger. Physics and Chemistry* (London: HMSO), **1**, pp. 199–209, 227–235.

Tolstoy, I. and Clay, C.S., 1966, *Ocean Acoustics: Theory and Experiment in Underwater Sound* (New York: McGraw-Hill), 293 pp.

UNESCO, Tenth Report of the Joint Panel on Oceanographic Tables and Standards, 1981, *UNESCO Technical Papers on Marine Science*, No. 36, 24 pp.

Chapter 3

Dopplick, T.G., 1972, Radiative heating of the global atmosphere, *Journal of Atmospheric Science*, **29**, 1278–1294.

Ivanoff, A., 1977, Oceanic absorption of solar energy, in *Modelling and Prediction of the Upper Layers of the Ocean*, ed. Kraus, E.B. (Oxford: Pergamon), Chapter 3.

Jerlov, N.G., 1976, *Marine Optics* (Amsterdam: Elsevier), 251 pp.

Kondratyev, K.Ya., 1969, *Radiation in the Atmosphere* (London: Academic Press), 912 pp.

Liou, K.N., 1980, *An Introduction to Atmospheric Radiation* (London: Academic Press).

London, J., 1957, *A Study of the Atmospheric Heat Balance: Final Report*, contract AF 19(122)-165, Department of Meteorology, New York University.

Manabe, S. and Strickler, R.F., 1964, Thermal equilibrium of the atmosphere with a convective adjustment, *Journal of Atmospheric Science*, **21**, 361–385.

Paltridge, G.W. and Platt, C.M.R., 1976, *Radiative Processes in Meteorology and Climatology* (Amsterdam: Elsevier), 318 pp.

Woods, J.D., Barkman, W. and Horch, A., 1984, Solar heating of the oceans: diurnal, seasonal and meridional variation, *Quarterly Journal of the Royal Meteorological Society*, **110**(465), 633–656.

Chapter 4

Ludlam, F.H., 1980, *Clouds and Storms* (University Park, PA: Pennsylvania State University Press), 405 pp.

Mason, B.J., 1971, *The Physics of Clouds* (Oxford: Clarendon Press), 671 pp.

Mason, B.J., 1975, *Clouds, Rain and Rainmaking* (Cambridge University Press), 189 pp.

Scorer, R.S., 1972, *Clouds of the World* (Newton Abbot, UK: David Charles), 176 pp.

Chapter 5

Broecker, W.S., 1981, Geochemical tracers and ocean circulation, in *Evolution of Physical Oceanography*, ed. Warren, B.A. and Wunsch, C. (Boston, MA: MIT Press), Chapter 15.

Brunt, D., 1932, Notes on radiation in the atmosphere, *Quarterly Journal of the Royal Meteorological Society*, **58**, 389–420.

Bunker, A.F., 1976, Computations of surface energy flux and annual air–sea interaction cycles in the North Atlantic, *Monthly Weather Review*, **104**, 1122–1140.

Lumb, F.E., 1964, The influence of clouds on hourly amounts of total solar radiation, *Quarterly Journal of the Royal Meteorological Society*, **90**, 43–56.

Newell, R.E., Kidson, J.W., Vincent, D.G. and Boar, G.J., 1972, *The General Circulation of the Tropical Atmosphere*, Vol. 1 (Boston, MA: MIT Press), 258 pp.

Oort, A.H. and von der Haar, T.H., 1976, On the observed annual cycle in the ocean–atmosphere heat balance over the northern hemisphere, *Journal of Physical Oceanography*, **6**, 781–800.

Priestley, C.H.B., 1959, *Turbulent Transfer in the Lower Atmosphere* (University of Chicago Press), 130 pp.

Sellers, W.D., 1965, *Physical Climatology* (University of Chicago Press), 272 pp.

Stommel, H., 1980, Asymmetry of interoceanic fresh-water and heat fluxes, *Proceedings of the National Academy of Science, USA*, **77**, 2377–2381.

Stommel, H. and Csanady, G.T., 1980, A relation between the T–S curve and global heat and atmospheric water transports, *Journal of Geophysical Research*, **85**, 495–501.

Trenberth, K.E., 1979, Mean annual poleward energy transports by the oceans in the Southern Hemisphere, *Dynamics of Atmosphere and Ocean*, **4**, 57–64.

Worthington, L.V., 1981, Water masses of the world ocean – a fine scale census, in *Evolution of Physical Oceanography*, ed. Warren, B.A. and Wunsch, C. (Boston, MA: MIT Press), Chapter 2.

Chapter 6

Burkov, V.A., Bulatov, R.P. and Neiman, V.G., 1973, Large scale features of water circulation in the world ocean, *Oceanology*, **13**, 325–332.

van der Hoven, J., 1957, Power spectrum of wind velocity in the frequency range from 0.0007 to 900 cycles per hour, *Journal of Meteorology*, **14**, 160–164.

Lorenz, E.N., 1969, *The Global Circulation of the Atmosphere* (London: Royal Meteorological Society).

Luyten, J.R. and Swallow, J.C., 1976, Equatorial undercurrents, *Deep Sea Research*, **23**(10), 999–1001.

Maury, M.F., 1853, *The Physical Geography of the Sea* (London: Sampson Low & Son), 274 pp.

Munk, W. and Wunsch, C., 1983a, Observing the ocean in the 1990s, *Philosophical Transactions of the Royal Society of London*, **A307**, 438–463.

Munk, W. and Wunsch, C., 1983b, The study of the ocean and the land surface from satellites, *Philosophical Transactions of the Royal Society of London*, **A309**, 243–464.

Newell, R.E. et al., 1972, *The General Circulation of the Tropical Atmosphere* (Boston, MA: MIT Press).

Reid, J.L., 1981, On the mid-depth circulation of the world ocean, in *Evolution of Physical Oceanography*, ed. Warren, B.A. and Wunsch, C. (Boston, MA: MIT Press), Chapter 3.

Rossby, H.T., Voorhis, A.D. and Webb, D., 1975, A quasi-Lagrangian study of mid-ocean variability using long-range SOFAR floats, *Journal of Marine Research*, **33**, 355–382.

Rossby, H.T., Riser, S.C. and Mariano, A.J., 1983, The western North Atlantic – a Lagrangian viewpoint, in *Eddies in Marine Science*, ed. Robinson, A.R. (New York: Springer-Verlag), 66–88.

Swallow, J.C., 1955, A neutral buoyancy float for measuring deep currents, *Deep Sea Research*, **3**, 74–81.

Swallow, J.C., 1971, The Aries current measurements in the western North Atlantic, *Philosophical Transactions of the Royal Society of London*, **A270**, 451–460.

The TWERLE Team, 1977, The TWERL experiment, *Bulletin of the American Meteorological Society*, **58**, 936–948.

Warren, B.A., 1981, Deep circulation in the world ocean, in *Evolution of Physical Oceanography*, ed. Warren, B.A. and Wunsch, C. (Boston, MA: MIT Press), Chapter 1.

Wunsch, C., 1981, Low frequency variability of the sea, in *Evolution of Physical Oceanography*, ed. Warren, B.A. and Wunsch, C. (Boston, MA: MIT Press), Chapter 11.

Chapter 7

Cheyney, R.E. and Marsh, J.G., 1981, Seasat altimeter observations of dynamic topography in the Gulf Stream region, *Journal of Geophysical Research*, C1, **86**, 473–483.

Ekman, V.W., 1905, On the influence of the earth's rotation on ocean currents, *Ark. F. Mat. Astron. Och. Fysik*, **2** (11), 1–53.

Findlater, J., 1971, *Geophysical Memoirs*, No. 115 (London: HMSO), 53 pp.

McWilliams, J.C., 1976, Maps from the mid-ocean dynamics experiment. Part 1. Geostrophic streamfunction, *Journal of Physical Oceanography*, **6**, 810–827.

Stommel, H., 1958, The abyssal circulation, *Deep Sea Research*, **5**, 80–82.

Stommel, H., 1966, *The Gulf Stream* (Cambridge University Press/University of California Press), 248 pp.

Stommel, H. and Arons, A.B., 1960, On the abyssal circulation of the world's ocean. 1. Stationary planetary flow patterns on a sphere, *Deep Sea Research*, **6**, 140–154.

Swallow, J.C. and Bruce, J.G., 1966, Current measurements off the Somali coast during the SW monsoon of 1964, *Deep Sea Research*, **13**, 861–888.

Chapter 8

Cartwright, D.E., Driver, J.S. and Tranter, J.E., 1977, Swell waves at Saint Helena related to distant storms, *Quarterly Journal of the Royal Meteorological Society*, **103**(438), 655–684.

Charnock, H., 1981, Air–sea interaction, in *Evolution of Physical Oceanography*, ed. Warren, B.A. and Wunsch, C. (Boston, MA: MIT Press), Chapter 17.

Flather, R.A., 1984, A numerical model investigation of the storm surge of

31 January and 1 February 1953 in the North Sea, *Quarterly Journal of the Royal Meteorological Society*, **110**(465), 591–612.

Kinsman, B., 1984, *Wind Waves: Their Generation and Propagation on the Ocean Surface* (New York: Dover Publications), 676 pp.

Pekeris, C.L. and Accad, Y., 1969, Solution of Laplace's equations for the M_2 tide in the world's ocean, *Philosophical Transactions of the Royal Society of London*, **A265**, 413–436.

Pugh, D.T., 1987, *Tides, Surges and Mean Sea Level: A Handbook for Engineers and Scientists* (Chichester: Wiley), 472 pp.

Thomson, W. (Lord Kelvin), 1879, On gravitational oscillations of rotating water, *Proceedings of the Royal Society, Edinburgh*, **10**, 92–100.

Chapter 9

Lorenz, E.N., 1967, *The Nature and Theory of the General Circulation of the Atmosphere* (World Meteorological Organization), 161 pp.

Newell, R.E., Vincent, D.G., Dopplick, T.G., Ferruzza, D. and Kidson, J.W., 1969, The energy balance of the global atmosphere, in *The Global Circulation of the Atmosphere*, ed. Corby, G.A. (London: Royal Meteorological Society), 42–90.

Newton, C.W., 1969, The role of extratropical disturbances in the global atmosphere, in *The Global Circulation of the Atmosphere*, ed. Corby, G.A. (London: Royal Meteorological Society), 137–158.

Palmen, E. and Newton, C.W., 1969, *Atmospheric Circulation Systems* (London: Academic Press), 603 pp.

Parker, C.E., 1971, Gulf Stream rings in the Sargasso Sea, *Deep Sea Research*, **18**(10), 981–993.

Chapter 10

Cane, M.A., 1983, Oceanographic events during El Niño, *Science*, **222**, 1195–1202.

Farmer, G., Wigley, T.M.L., Jones, P.D. and Salmon, M., 1989, *Documenting and Explaining Recent Global Mean Temperature Changes*, Climate Research Unit, Norwich, UK. Final Report to NERC, UK.

Gates, W.L., 1976, Modelling the ice-age climate, *Science*, **191**, 1138–1144.

Ghil, M., 1980, Internal climatic mechanisms participating in glaciation cycles, in *Climatic Variations and Variability: Facts and Theories*, ed. Berger, A., NATO Advanced Study Series (Dordrecht: Reidel), 539–558.

Houghton, J.T., Jenkins, G.J. and Ephraums, J.J. (eds), 1990, *Climate Change: The IPCC Scientific Assessment* (Cambridge University Press), 365 pp.

Imbrie, J., 1980, Time dependent models of the climate response to orbital variations, in *Climatic Variations and Variability: Facts and Theories*, ed. Berger, A., NATO Advanced Study Series (Dordrecht: Reidel).

JASIN, 1983, Results of the Royal Society Joint Air–Sea Interaction Project

(JASIN), *Philosophical Transactions of the Royal Society of London*, **A308**, 221–449.

Lovelock, J., 1988, *The Ages of Gaia: A Biography of our Living Earth* (Oxford: Oxford University Press), 252 pp.

Mitchell, J.M., 1975, A note on solar variability and volcanic variability, in *Physical Basis of Climate and Climate Modelling* (World Meteorological Organization), Garp 16.

Munk, W., 1989, Global ocean warming: detection by long-path acoustic travel times, *Oceanography*, **2**(2), 40–41.

Munk, W. and Wunsch, C., 1983a, Observing the ocean in the 1990s, *Philosophical Transactions of the Royal Society of London*, **A307**, 438–463.

Munk, W. and Wunsch, C. 1983b, The study of the ocean and the land surface from satellites, *Philosophical Transactions of the Royal Society of London*, **A309**, 243–464.

Namias, J., 1971, The 1968–69 winter as an outgrowth of sea and air coupling during antecedent seasons, *Journal of Physical Oceanography*, **1**, 65–81.

Namias, J., 1973, Hurricane Agnes – an event shaped by large-scale air–sea systems generated during antecedent months, *Quarterly Journal of the Royal Meteorological Society*, **99**(421), 506–519.

Philander, S.G.H., 1990, *El Niño, La Niña and the Southern Oscillation* (New York: Academic Press), 293 pp.

Pittock, A.B., Frakes, L.A., Jenssen, D., Peterson, J.A. and Zillman, J.W. (eds), 1978, *Climate Change and Variability: A Southern Ocean Perspective* (Cambridge University Press), 455 pp.

Smagorinsky, J., 1953, The dynamical influence of large-scale heat sources and sinks on quasi-stationary mean motions of the atmosphere, *Quarterly Journal of the Royal Meteorological Society*, **79**(341), 342–366.

Trenberth, K.E., 1992, *Climate System Modelling* (Cambridge University Press), 788 pp.

WMO, Garp Publication Series 16, 1975, *The Physical Basis of Climate and Climate Modelling*.

WMO, Garp Publication Series 24, 1981, *Scientific Results of the Air Mass Transformation Experiment*, 236 pp.

Figure Sources

Figure 1.1 Allen, C.W., 1958, *Quarterly Journal of the Royal Meteorological Society*, **84**: page 311, figure 3. Reproduced with permission.

Figure 1.2 Pittock, A.B. *et al.* eds, 1978, *Climate Change and Variability: A Southern Perspective*, Cambridge University Press: page 10, figure 2.1.1. Reproduced by permission of Cambridge University Press.

Figure 1.3 Hess, S.L., 1959, *Introduction to Theoretical Meteorology*, Holt, Rinehart & Winston: page 132, figure 9.1.

Figure 1.4 Jacobowitz, H. *et al.*, 1979, *Journal of Atmospheric Science*, **36**: page 506, figure 6. Reproduced by permission of American Meteorological Society.

Figure 1.6 Pittock, A.B. *et al.* eds, 1978, *Climate Change and Variability: A Southern Perspective*, Cambridge University Press: page 12, figure 2.1.2. Reproduced by permission of Cambridge University Press.

Figure 1.7 Jacobowitz, H. *et al.*, 1979, *Journal of Atmospheric Science*, **36**: page 506, figure 7. Reproduced by permission of American Meteorological Society.

Figure 1.8 McLellan, H.J., 1965, *Elements of Physical Oceanography*, Pergamon Press: page 5, figure 1.2.

Figure 1.9 McLellan, H.J., 1965, *Elements of Physical Oceanography*, Pergamon Press: page 4, figure 1.1.

Figure 2.1 Friedrich, H. and Levitus, S., 1972, *Journal of Physical Oceanography*, **2**: page 516, figure 4. Reproduced by permission of American Meteorological Society.

Figure 2.2 Tchernia, P., 1980, *Descriptive Regional Oceanography*, Pergamon Press: page 162, figure 5.40.

Figure 2.3 Tchernia, P., 1980, *Descriptive Regional Oceanography*, Pergamon Press: page 35, figure 3.4.

Figure 2.4 Tolstoy, I. and Clay, C.S., 1966, *Ocean Acoustics: Theory and Experiment in Underwater Sound*, McGraw-Hill: page 7, figure 1.2.

Figure 2.6 Tchernia, P., 1980, *Descriptive Regional Oceanography*, Pergamon Press: page 30, figure 3.2.

Figure 2.9 Stull, R.B., 1973, *Journal of Atmospheric Science*, **30**: page 1097, figure 5. Reproduced by permission of American Meteorological Society.

Figure 3.1   Goody, R.M. and Walker, J.C.G., 1972, *Atmospheres*, Prentice-Hall: page 25, figure 2.2.

Figure 3.2   Goody, R.M., 1964, *Atmospheric Radiation*, Oxford University Press: page 4, figure 1.1. Reproduced by permission of Oxford University Press.

Figure 3.3   Ivanoff, A., 1977, Ocean absorption of solar energy in *Modelling and Prediction of the Upper Layers of the Ocean*, ed. Kraus, E.B.: (Oxford: Pergamon): page 49, figure 5.3.

Figure 3.5   Adetunji, J. *et al.*, 1979, *Weather*, **30**: pages 434 and 435, figures 4 and 5. Reproduced by permission of the Royal Meteorological Society.

Figure 3.6   Schmetz *et al.*, 1983, *Results of the Royal Society Joint Air–Sea Interaction Project (JASIN)*, The Royal Society: page 380, figure 2. Reproduced with permission

Figure 3.7   Wallace, J.M. and Hobbs, P.V., 1977, *Atmospheric Science*, Academic Press: page 23, figure 1.8.

Figure 3.9   Manabe, S. and Strickler, R.F., 1964, *Journal of Atmospheric Science*, **21**: page 371, figure 6a. Reproduced by permission of American Meteorological Society.

Figure 3.10   Henderson-Sellers, A. and Wilson, M.F., 1983, *Philosophical Transactions of the Royal Society*, **A309**: page 286, figure 1. Reproduced with permission

Figure 3.11   Dopplick, T.G., 1972, *Journal of Atmospheric Science*, **29**: pages 1291 and 1292, figures 21 and 23. Reproduced by permission of American Meteorological Society.

Figure 3.12   Jerlov, N.G., 1976, *Marine Optics*, Elsevier: page 140, figure 7.4. Reproduced by permission of Elsevier.

Figure 3.13   Woods, J.D. *et al.*, 1984, *Quarterly Journal of the Royal Meteorological Society*, **110**(465): page 640, figure 3a. Reproduced with permission.

Figure 3.14   Pickard, G.L. and Emery, W.J. Descriptive Physical Oceanography (5th Edition), Pergamon Press: page 45, figure 4.6. Reproduced by permission of Butterworth-Heinemann Ltd.

Figure 4.3   Lorenz, E.N., 1967, *The Nature and Theory of the General Circulation of the Atmosphere*, World Meteorological Organization: 218 TP 115, pages 42 and 43, figures 14 and 16. Reproduced by permission of World Meteorological Organisation.

Figure 4.6   Mason, B.J., 1975, *Clouds, Rain and Rainmaking*, Cambridge University Press: page 21, figure 4. Reproduced by permission of Cambridge University Press.

Figure 4.7   Slinn, N. and George, W., 1975, *Atmospheric Environment*, **9**: page 763, figure 1. Reproduced by permission of Pergamon Press.

Figure 4.8   Wallace, J.M. and Hobbs, P.V., 1977, *Atmospheric Science*, Academic Press: page 171, figure 4.16.

Figure 4.9   Wallace, J.M. and Hobbs, P.V., 1977, *Atmospheric Science*, Academic Press: page 73, figure 2.7.

Figure 4.10   Wallace, J.M. and Hobbs, P.V., 1977, *Atmospheric Science*, Academic Press: page 255, figure 5.32.

Figure 4.11 Mason, B.J., 1975, *Clouds, Rain and Rainmaking*, Cambridge University Press: page 94, figure 24. Reproduced by permission of Cambridge University Press.

Figure 5.1 Priestley, C.H.B., 1959, *Turbulent Transfer in the Lower Atmosphere*, University of Chicago Press: page 67, figure 18.
Figure 5.2 Bunker, A.F., 1976, *Monthly Weather Review*, **104**: pages 1130 and 1131, figures 4a, 4b, 4c and 4d. Reproduced by permission of American Meteorological Society.
Figure 5.3 Oort, A.H. and von der Haar, T.H., 1976, *Journal of Physical Oceanography*, **6**: pages 789–795, figures 5, 8, 9 and 13. Reproduced by permission of American Meteorological Society.
Figure 5.4 Stommel, H., 1980, *Proceedings of the National Academy of Science, USA*, **77**: page 2378, figure 1. Reproduced by permission of American Meteorological Society.
Figure 5.5 Oort, A.H. and von der Haar, T.H., 1976, *Journal of Physical Oceanography*, **6**: page 783, figure 1. Reproduced by permission of American Meteorological Society.
Figure 5.6 Carrissimo, B.C., Oort, A.H. and von der Haar, T.H., 1985, *Journal of Physical Oceanography*, **15**: page 85, figure 4. Reproduced by permission of American Meteorological Society.
Figure 5.7 Schmitt, R.W. and Wijffels, S.E., 1993, In *Interactions between Global Climate Sub-systems: The Legacy of Hann*, ed. McBean, G.A. and Hantel, M., **75**, Geophysical Monographs: page 4, figure 2.
Figure 5.8 Wijffels, S.E., Schmitt, R., Bryden, H. and Stigebrandt, A., 1992, *Journal of Physical Oceanography*, **22**: page 158, figure 2. Reproduced by permission of American Meteorological Society.
Figure 5.9 Tchernia, P., 1980, *Descriptive Regional Oceanography*, Pergamon Marine Series 3: page 38, figure 3.5.
Figure 5.10 Dietrich, G. *et al.*, 1980, *General Oceanography: An Introduction*, Wiley Interscience: page 197, figure 78. Reproduced by permission of Interscience/Wiley.
Figure 5.11 Gross, G.N., 1987, *Oceanography: A View of the Earth*, Prentice-Hall, page 449, figure 7.15.
Figure 5.12 Tchernia, P., 1980, *Descriptive Regional Oceanography*, Pergamon Marine Series 3: page 157, figure 5.34.
Figure 5.13 Worthington, L.V., 1981, In *Evolution of Physical Oceanography*, ed. Warren, B.A. and Wunsch, C., MIT Press: chapter 2, page 47, figure 2.2. Reproduced by permission.
Figure 5.15 Broecker, W.S. and Peng, T.H., 1982, *Tracers in the Sea*, Eldigo Press: page 409, figure 8.15. Reproduced by permission of Eldigo Press.

Figure 6.2 Rossby, H.T. *et al.*, 1983, In *Eddies in Marine Science*, ed. Robinson, A.R., Springer-Verlag: pages 68 and 70, figures 1 and 3.
Figure 6.3 Luyten, J.R. and Swallow, J.C., 1976, *Deep Sea Research*, **23**(10): page 1001, figure 1. Reproduced by permission of Pergamon Press.

Figure 6.5 van der Hoven, J., 1957, *Journal of Meteorology*, **14**: page 161, figure 1. Reproduced by permission of American Meteorological Society.
Figure 6.6 Rhines, P.B., 1971, *Deep Sea Research*, **18**: pages 21–26, figure 1. Reproduced by permission of Pergamon Press.
Figure 6.7 Lorenz, E.N., 1969, In *The Global Circulation of the Atmosphere*, ed. Corby, G.A., Royal Meteorological Society: page 11, figure 4. Reproduced with permission.
Figure 6.8 The TWERLE Team, 1977, *Bulletin of the American Meteorological Society*, **58**: page 942, figure 6a. Reproduced by permission of American Meteorological Society.
Figure 6.9 Rossby, H.T. *et al.*, 1975, *Journal of Marine Research*, **33**: pages 366 and 369, figures 6 and 10. Reproduced by permission of Science Press (Sears Foundation for Marine Research).
Figure 6.10 Newell, R.E. *et al.*, 1969, In *The Global Circulation of the Atmosphere*, ed. Corby, G.A., Royal Meteorological Society: page 69, figures 14a and b. Reproduced with permission.
Figure 6.11 Newell, R.E. *et al.*, 1969, In *The Global Circulation of the Atmosphere*, ed. Corby, G.A., Royal Meteorological Society: page 65, figures 9a and b. Reproduced with permission.
Figure 6.12 Gross, G.M., 1987, *Oceanography: A View of the Earth*, Prentice Hall, page 132, figure 6.11.
Figure 6.13 (a) NRDC, 1946, *Summary Technical Report*, Office of Naval Research, Washington, DC: page 68, figure 15.
Figure 6.13 (b) and (c) Gross, G.M., 1987, *Oceanography: A View of the Earth*, Prentice-Hall: page 180, figure 7.5.

Figure 7.4 Neumann, G. and Pierson, W., 1966, *Principles of Physical Oceanography*, Prentice-Hall: page 159, figure 7.19.
Figure 7.7 Nicholls, S. *et al.*, 1983, In *Results of the Royal Society Joint Air–Sea Interaction Project (JASIN)*, The Royal Society: page 299, figure 3. Reproduced with permission.
Figure 7.8 Cheney, R.E. and Marsh, J.G., 1981, *Journal of Geophysical Research*, C1, **86**: page 480, figure 9. Reproduced with permission.
Figure 7.9 McWilliams, J.C., 1976, *Journal of Physical Oceanography*, **6**: page 813, figure 1 Reproduced by permission of American Meteorological Society.
Figure 7.12 Stommel, H., 1966, *The Gulf Stream*, Cambridge University Press/University of California Press: page 92, figure 58.
Figure 7.14 Stommel, H., 1958, *Deep Sea Research*, **5**: page 81, figure 1. Reproduced by permission of Pergamon Press.
Figure 7.17 Rowe, M.A., 1985, University of Southampton: PhD thesis.
Figure 7.20 (a) Findlater, J., 1971, *Geophysical Memoirs*, **115**: page 2, figure 1. Crown ©. Reproduced with the permission of the Controller of Her Majesty's Stationery Office.
Figure 7.20 (b) Swallow, J.C. and Bruce, J.G., 1966, *Deep Sea Research*, **13**: page 866, figure 2. Reproduced by permission of Pergamon Press.

Figure 8.1   Kinsman, B., 1984, *Wind Waves: Their Generation and Propagation on the Ocean Surface*, Prentice-Hall, page 23, figure 1.2.1.

Figure 8.2   Gross, G.M., 1987, *Oceanography: A View of the Earth*, Prentice-Hall: page 208, figure 8.4.

Figure 8.3   Charnock, H., 1981, In *Evolution of Physical Oceanography*, ed. Warren, B.A. and Wunsch, C., MIT Press: chapter 17, page 493, figure 17.6. Reproduced with permission.

Figure 8.4   Revault-d'Allones, M. and Caulliez, G., 1980, In *Marine Turbulence*, ed. Nihoul, J.C.J., Elsevier: page 110, figure 5.

Figure 8.5   Charnock, H., 1981, In *Evolution of Physical Oceanography*, ed. Warren, B.A. and Wunsch, C., MIT Press: chapter 17, page 493, figure 17.8. Reproduced with permission.

Figure 8.6   Cartwright, D.E. *et al.*, 1977, *Quarterly Journal of the Royal Meteorological Society*, **103**(438): page 664, figures 4a and b. Reproduced with permission.

Figure 8.7   Gill, A.E., 1982, *Atmosphere–Ocean Dynamics*, Academic Press: page 380, figure 10.3.

Figure 8.9   Dietrich, G. *et al.*, 1980, *General Oceanography*, Wiley Interscience: page 421, figure 180. Reproduced by permission of Interscience/Wiley.

Figure 8.10   Schwiderski, E.W., 1979, *Global Ocean Tides, part II: the Semi-diurnal Principal Lunar Tide M_2. Atlas of Tidal Charts and Maps*, United States Naval Surface Weapons Center, *Technical Report Center*, NSWC TR-79-414.

Figure 8.11   Flather, R.A., 1984, *Quarterly Journal of the Royal Meteorological Society*, **110**(465): page 594, figure 1. Reproduced with permission.

Figure 8.12   Flather, R.A., 1984, *Quarterly Journal of the Royal Meteorological Society*, **110**(465): pages 605–608, figure 5. Reproduced with permission.

Figure 9.7   Palmen, E. and Newton, C.W., 1969, *Atmospheric Circulation Systems*, Academic Press: page 176, figure 7.4.

Figure 9.9   Palmen, E. and Newton, C.W., 1969, *Atmospheric Circulation Systems*, Academic Press: page 310, figure 10.20.

Figure 9.10   (a) and (b) Richardson, P.L., 1983, In *Eddies in Marine Science*, ed. Robinson, A.R., Springer-Verlag: pages 20 and 30, figures 1 and 6.

Figure 9.11   Newell, R.E. *et al.*, 1969, In *The Global Circulation of the Atmosphere*, ed. Corby, G.A., Royal Meteorological Society: page 63, figure 7. Reproduced with permission.

Figure 9.12   Newton, C.W., 1969, In *The Global Circulation of the Atmosphere*, ed. Corby, G.A., Royal Meteorological Society: page 138, figure 1. Reproduced with permission.

Figure 9.13   Lorenz, E.N., 1967, *The Nature and Theory of the General Circulation of the Atmosphere*, World Meteorological Organization: page 82, figure 40. Reproduced by permission of Cambridge University Press.

Figure 9.14   Wells, N.C., Gould, W.J. and Kemp, A.E.S., 1996, In *Oceanography – An Illustrated Guide*, ed. Summerhayes, C.P. and Thorpe,

S.A.T., Manson Publishing: page 52, figure 3.18. Reproduced by permission of Manson Publishing

Figure 10.2   WMO, 1981, *Garp Publication Series*, **24**: page 167, figure 6.1. Reproduced by permission of World Meteorological Organization.
Figure 10.3   Guymer, T.H. *et al.*, 1983, *Philosophical Transactions of the Royal Society of London*, **A308**: page 265, figures 10 and 11. Reproduced with permission.
Figure 10.4   Smagorinsky, J., 1953, *Quarterly Journal of the Royal Meteorological Society*, **79**(341): page 362, figure 7. Reproduced with permission.
Figure 10.5   Namias, J., 1973, *Quarterly Journal of the Royal Meteorological Society*, **99**(421): pages 509 and 517, figures 2 and 10. Reproduced with permission.
Figure 10.7   Philander, G., 1983, *Nature*, **302**: page 297, figure 3. Reproduced with permission.
Figure 10.8   Wells, N.C., Gould, W.J. and Kemp, A.E.S., 1996, In *Oceanography – An Illustrated Guide*, ed. Summerhayes, C.P. and Thorpe, S.A.T., Manson Publishing: page 54, figure 3.22. Reproduced by permission of Manson Publishing.
Figure 10.9   Folland, C.K., Tarl, T.R. and Vinnikov, K.Y.A., 1990, In *Climate Change: The IPCC Scientific Assessment*, ed. Houghton, J.T. *et al.*, Cambridge University Press: page 211, figure 7.9. Reproduced by permission of Cambridge University Press.
Figure 10.10   Lazier, J., 1980, *Atmosphere and Ocean*, **18**: page 230, figure 2. Reproduced with permission of the Canadian Meteorological and Oceanographic Society.
Figure 10.11   Folland, C.K., Tarl, T.R. and Vinnikov, K.Y.A., 1990, In *Climate Change: The IPCC Scientific Assessment*, ed. Houghton, J.T. *et al.*, Cambridge University Press: page 202, figure 7.1. Reproduced by permission of Cambridge University Press.
Figure 10.12   Wells, N.C., Gould, W.J. and Kemp, A.E.S., 1996, In *Oceanography – An Illustrated Guide*, ed. Summerhayes, C.P. and Thorpe, S.A.T., Manson Publishing: page 42, figure 3.1. Reproduced by permission of Manson Publishing.
Figure 10.13   Wunsch, C. and Gaposchkin, E.M., 1980, *Reviews of Geophysics and Space Physics*, **18**: page 734, figure 10. Reproduced with permission.

Appendices

A Standard International (SI) Units

Quantity	Symbol	Name	Definition
Force	N	Newton	$kg\,ms^{-2}$
Pressure	Pa	Pascal	$kg\,m^{-1}\,s^{-2}$
Energy/Work	J	Joule	$kg\,m^2\,s^{-2}$
Power	W	Watt	$kg\,m^2\,s^{-3}$
Temperature	K	Kelvin	
Celsius Temperature	°C	degree Celsius	K-273.15

B SI Unit Prefixes

Name	Symbol	Multiplying factor 10^N, where N is given below.
peta	P	15
tera	T	12
giga	G	9
mega	M	6
kilo	k	3
hecto	h	2
deca	da	1
deci	d	−1
centi	c	−2
milli	m	−3
micro	μ	−6
nano	n	−9
pico	p	−12
femto	f	−15
atto	a	−18

Index